Lecture Notes in Computer Science 11083

Commenced Publication in 1973
Founding and Former Series Editors:
Gerhard Goos, Juris Hartmanis, and Jan van Leeuwen

More information about this series at http://www.springer.com/series/7412

Guorong Wu · Islem Rekik
Markus D. Schirmer · Ai Wern Chung
Brent Munsell (Eds.)

Connectomics in NeuroImaging

Second International Workshop, CNI 2018
Held in Conjunction with MICCAI 2018
Granada, Spain, September 20, 2018
Proceedings

Springer

Editors
Guorong Wu
University of North Carolina at Chapel Hill
Chapel Hill, NC
USA

Ai Wern Chung
Harvard Medical School
Boston, MA
USA

Islem Rekik
University of Dundee
Dundee
UK

Brent Munsell
College of Charleston
Charleston, SC
USA

Markus D. Schirmer
Harvard Medical School
Boston, MA
USA

ISSN 0302-9743   ISSN 1611-3349   (electronic)
Lecture Notes in Computer Science
ISBN 978-3-030-00754-6   ISBN 978-3-030-00755-3   (eBook)
https://doi.org/10.1007/978-3-030-00755-3

Library of Congress Control Number: 2018954661

LNCS Sublibrary: SL6 – Image Processing, Computer Vision, Pattern Recognition, and Graphics

This Springer imprint is published by the registered company Springer Nature Switzerland AG
The registered company address is: Gewerbestrasse 11, 6330 Cham, Switzerland

Preface

The 2nd International Workshop on Connectomics in NeuroImaging (CNI 2018) was held in Granada, Spain, on September 20th, 2018, in conjunction with the 21st International Conference on Medical Image Computing and Computer Assisted Intervention (MICCAI).

Connectomics is the study of whole brain maps of connectivity, commonly referred to as the brain connectome. It focuses on quantifying, visualizing, and understanding brain network organization, and includes applications in neuroimaging. The primary academic objective of the CNI workshop is to bring together computational researchers (computer scientists, data scientists, computational neuroscientists) to discuss new advancements in network construction, analysis, and visualization techniques in connectomics and their use in clinical diagnosis and group comparison studies. The secondary academic objective is to attract neuroscientists and clinicians to show recent methodological advancements in connectomics, and how they are successfully applied in various neuroimaging applications. CNI 2018 was held as a single-track workshop that included three keynote speakers (Gustavo Deco, Martijn van den Heuvel, and Dafnis Batalle), oral paper presentations, poster sessions, and software demonstrations.

The quality of submissions to our workshop was very high. Authors were asked to submit papers of 8–10 pages in length for review. A total of 20 papers were submitted to the workshop in response to our call for papers. Each of the 20 papers underwent a rigorous double-blind peer-review process, with each paper being reviewed by at least two reviewers from the Program Committee, composed of 28 well-known experts in the field of connectomics. Based on the reviewing scores and critiques, the best 15 papers were accepted for presentation at the workshop, and chosen to be included in this Springer LNCS volume. The large variety of connectomics techniques applied in neuroimaging applications were well represented at the CNI 2018 workshop.

We are grateful to the Program Committee for reviewing the submitted papers and giving constructive comments and critiques, to the authors for submitting high-quality papers, to the presenters for excellent presentations, and to all the CNI 2018 attendees who came to Granada from all around the world.

September 2018

Guorong Wu
Islem Rekik
Markus Schirmer
Ai Wern Chung
Brent Munsell

Organization

Program Committee

Carl Westin	Harvard Medical School, USA
Todd Constable	Yale University, USA
Martin Styner	University of North Carolina, USA
Barbara Marebwa	Medical University of South Carolina, USA
Emilie Mckinnon	Medical University of South Carolina, USA
Yong Fan	University of Pennsylvania, USA
Pierre Besson	Aix-Marseille Université, France
Yangming Ou	Harvard Medical School, USA
Johnathan O'Muircheartaigh	King's College London, UK
Gareth Ball	Murdoch Children's Research Institute, Australia
Gloria Menegaz	University of Verona, Italy
Silvina Ferradal	Harvard Medical School, USA
Pratik Mukherjee	University of California San Francisco, USA
Moo Chung	University of Wisconsin at Madison, USA
Zhifeng Kou	Wayne State University, USA
Archana Venkataraman	Johns Hopkins University, USA
Bernard Ng	University of British Columbia, Canada
Li Shen	University of Pennsylvania, USA
Boris Gutman	Illinois Institute of Technology, USA
Yong He	Beijing Normal University, China
Sylvain Bouix	Harvard Medical School, USA
Renaud Lopes	University of Lille Nord de France, France
Wei Gao	Cedars-Sinai Hospital, USA
Eran Dayan	University of North Carolina, USA
Ziwen Peng	Southern China Normal University, China
Mayssa Soussia	École Nationale d'Ingénieurs de Tunis, Tunisia
Oualid Benkarim	Universitat Pompeu Fabra, Spain
YingYing Zhu	University of North Carolina, USA

Contents

Towards Ultra-High Resolution 3D Reconstruction of a Whole Rat Brain from 3D-PLI Data

Sharib Ali[1]([✉]), Martin Schober[2], Philipp Schlömer[2], Katrin Amunts[2,3], Markus Axer[2], and Karl Rohr[1]

[1] Department of Bioinformatics, Biomedical Computer Vision Group, BIOQUANT, IPMB, DKFZ, University of Heidelberg, Heidelberg, Germany
sharib.ali@bioquant.uni-heidelberg.de
[2] Research Centre Jülich, Institute of Neuroscience and Medicine 1, Jülich, Germany
[3] Cécile and Oskar Vogt Institute of Brain Research,
Heinrich Heine University Düsseldorf, University Hospital Düsseldorf,
Düsseldorf, Germany

Abstract. 3D reconstruction of the fiber connectivity of the rat brain at microscopic scale enables gaining detailed insight about the complex structural organization of the brain. We introduce a new method for registration and 3D reconstruction of high- and ultra-high resolution (64 μm and 1.3 μm pixel size) histological images of a Wistar rat brain acquired by 3D polarized light imaging (3D-PLI). Our method exploits multi-scale and multi-modal 3D-PLI data up to cellular resolution. We propose a new feature transform-based similarity measure and a weighted regularization scheme for accurate and robust non-rigid registration. To transform the 1.3 μm ultra-high resolution data to the reference block-face images a feature-based registration method followed by a non-rigid registration is proposed. Our approach has been successfully applied to 278 histological sections of a rat brain and the performance has been quantitatively evaluated using manually placed landmarks by an expert.

1 Introduction

Studying the brain fiber architecture and their functionality, like that of the rat brain, is important for understanding complex human brain organization. Conventional imaging methods include electron microscopy (EM), optical microscopy (OM), and diffusion magnetic resonance imaging (D-MRI). While D-MRI is limited in resolution, EM and OM often require some selective staining procedure of histological brain sections to reveal fiber connectivity. Recent advances in 3D polarized light imaging (3D-PLI, a specialized OM technique that utilizes the birefringence of nerve fibers) allows acquiring high- and ultra-high resolution images of fibrous brain tissues [5]. In addition, information about 3D fiber orientation can be obtained without staining. 3D-PLI data consists of different image modalities (Fig. 1, right): Transmittance map representing the

© Springer Nature Switzerland AG 2018
G. Wu et al. (Eds.): CNI 2018, LNCS 11083, pp. 1–10, 2018.
https://doi.org/10.1007/978-3-030-00755-3_1

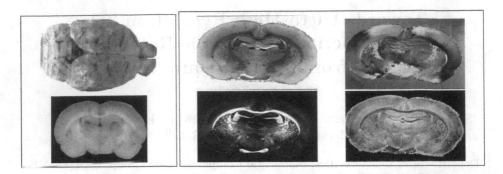

Fig. 1. Rat brain data. Left box: Reference blockface with 3D blockface volume (top) and mid-section (bottom). Right box: original ultra-high resolution (1.3 μm) 3D-PLI data comprising transmittance (top left), retardation (bottom left), direction (top-right), and inclination (bottom-right) maps. Images are scaled for better visualization.

extinction of polarized light when passing through the brain tissue, Retardation map showing the tissue's (fiber's) birefringence, as well as direction and inclination maps representing the local 3D fiber orientation. Blockface images are acquired during the sectioning procedure (Fig. 1, left) and constitute undistorted reference images for the acquired histological sections.

During the sectioning and mounting process brain tissue undergoes strong distortions. Thus, spatial coherence between sections is lost and hence image registration becomes an inevitable task. In previous work, 3D reconstruction of histological sections of the rat brain (e.g., [9–11]) was performed using rigid or affine registration (e.g. [4,11]), which is generally not sufficient to cope with deformations in histological sections as mentioned in [9]. [10] used affine registration with subsequent diffeomorphic non-rigid registration employing mutual information. Compared to traditional histological data, 3D-PLI relies on unstained cryo-sections and is acquired at very different resolutions. This poses different challenges compared to traditional histological data. In previous work on the registration and 3D reconstruction of 3D-PLI data, high-resolution images (64μm pixel size) were used in [1,13] and ultra-high resolution (1.3 μm pixel size) images in [2]. However, in [2] only rigid registration of the ultra-high resolution data to unregistered high-resolution images was performed, and the human brain was considered but not rat brain. [1,3] used high-resolution human brain sections (64 μm pixel size) for registration to reference blockface data of the same resolution. Note that human brain sections typically cover larger areas, contain more prominent structures, and include less image noise compared to the rat brain. Thus, registration of 3D-PLI data of the rat brain is more difficult. In [13], high-resolution 3D-PLI data (64 μm) of the rat brain was first registered to the blockface data of same resolution and then transformed to a reference Waxholm space. However, in contrast to [13], we register high-resolution images with a section thickness of 60 μm to blockface images of 15.5 μm pixel resolution, which is more challenging due to the large scale difference. Also, we subsequently register ultra-high resolution images (1.3 μm) first to the registered high-resolution images (15.5 μm after scaling) and then to upscaled reference blockface images at 1.3 μ m resolution using non-rigid registration (each image section has a size of about 15000 × 12000 pixels). In addition, whereas in [13] B-splines and a fluid model were used, respectively, we here use a more realistic deformation model based on Gaussian non-rigid body splines (GEBS) for non-rigid registration. None of the previous work provided a complete framework for ultra-high resolution 3D reconstruction of the rat brain from 3D-PLI.

In this contribution, we introduce a new method for multi-scale (both high- and ultra-high resolution data) and multi-modal registration of histological rat brain sections from 3D-PLI. The main contributions are: (1) registration of 3D-PLI data with three different spatial resolutions (1.3 μm, 15.5 μm, and 64 μm pixel size), (2) correlation transform-based similarity metric for efficient and robust rigid registration, (3) introduction of a feature transform-based similarity metric and weighted regularization for non-rigid registration using a physically-based deformation model, (4) robust feature-based registration, and (5) a complete pipeline for 3D reconstruction.

2   Method

Our approach for 3D reconstruction of both high- and ultra-high resolution 3D-PLI data of the rat brain consists of several steps. High resolution images are first registered to their corresponding reference blockface images using rigid and non-rigid registration. Ultra-high resolution images are then registered both rigidly and non-rigidly to the corresponding sections of the reference blockface images.

2.1   Registration of High-Resolution 3D-PLI Data

To coherently align the high-resolution 3D-PLI data with the reference block-face images several registration steps are required. The high-resolution data is

Fig. 2. Pre-processing of high-resolution 3D-PLI data. Left: Original image data, middle: segmented and scaled image, and right: COM alignment with blockface image.

first coarsely registered using center-of-mass alignment, rigid registration, and then non-rigid registration using GEBS [8] in conjunction with a novel feature transform-based similarity measure and a weighted quadratic regularization.

Data Preparation and Coarse Registration. High resolution 3D-PLI sections of the rat brain are segmented from the original image data (see Fig. 2, left) as in [1]. For initial alignment we perform a scaling transformation for high-resolution images (64 μm) and then align their center-of-mass (COM) with that of the reference blockface images (15.5 μm, see Fig. 2, right).

We use a parametric registration model for coarse registration of 3D-PLI data. Let $g_1(\mathbf{x})$ and $g_2(\mathbf{x})$ with $\mathbf{x} = (x, y) : \Omega \rightarrow \mathbb{R}, \Omega \in \mathbb{R}^2$, be the reference blockface and the PLI image, respectively, and $\mathcal{T}(\mathbf{x} \mid \theta)$ be the transformation with the parameter vector θ to be estimated. Then, the goal is to minimize the objective function ψ to obtain the optimal $\hat{\theta}$:

$$\hat{\theta} = \underset{\theta}{\arg\min} \, \psi \left(g_1(\mathbf{x}), \, g_2\big(\mathcal{T}(\mathbf{x} \mid \theta)\big) \right). \tag{1}$$

We use a spline-based multi-resolution scheme for rigid registration based on [14]. In contrast to [14], where the sum of squared intensity differences (SSD) was employed, we propose using a correlation transform (CoT) of the image to deal with multi-modal data (see Fig. 1). Let $P_{\mathbf{x}}$ be a patch of size 7×7 pixels centered at $\mathbf{x}$, then the CoT is given by

$$\tilde{g}(\mathbf{x}) = (g(\mathbf{x}_k) - \mu) / (\sigma + \epsilon), \quad \text{with } \mathbf{x}_k \in P_{\mathbf{x}}, \tag{2}$$

where μ and σ are the mean intensity and standard deviation, respectively within $P_{\mathbf{x}}$ and $\epsilon = 0.001$. For ψ in (1) we use the SSD between the computed CoT values for the blockface image $\tilde{g}_1$ and the high-resolution image $\tilde{g}_2$: $\psi(\theta) = \sum_{\mathbf{x} \in \Omega} \left(\tilde{g}_1(\mathbf{x}) - \tilde{g}_2\big(\mathcal{T}(\mathbf{x} \mid \theta)\big) \right)^2$. We minimize Eq. (1) using Levenberg-Marquardt optimization.

Non-rigid Registration. Non-linear distortions are often present in 3D-PLI data due to the cutting and mounting procedure. In addition, local deformations are introduced because of time delays between mounting and data acquisition. Since these deformations are the result of physical phenomena, a suitable physical deformation model should be used for non-rigid registration. In our approach, we use Gaussian elastic body splines (GEBS) which represent an analytic solution of the Navier equation from linear elasticity theory [7]: $\mu \Delta \mathbf{u} + (\lambda + \mu) \nabla (\operatorname{div} \mathbf{u}) + \mathbf{f} = \mathbf{0}$, where λ and $\mu > 0$ are the Lamé constants and $\mathbf{u}$ is the deformation field under Gaussian forces $\mathbf{f}$, and which has been derived in [8]. In [15], an intensity-based registration approach using GEBS was described, which, however is not suitable for multi-modal 3D-PLI data. Using a CoT-based similarity measure for non-rigid registration has disadvantages (see the red arrows in Fig. 3 which indicate that structure and intensity invariance are not well preserved). In this contribution, we introduce a feature transform-based

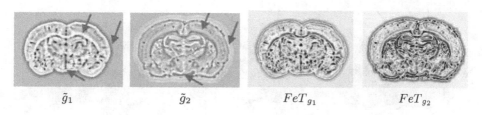

$\tilde{g}_1$ $\tilde{g}_2$ FeT_{g_1} FeT_{g_2}

Fig. 3. Correlation transform ($\tilde{g}$) and feature-transform (FeT) images of the multi-modal reference blockface (g_1) and 3D-PLI (g_2) data (also refer to Fig. 1).

(FeT) similarity measure, and a Gaussian weighted quadratic regularization. FeT better preserves the structure and intensity invariance and is thus better suited for non-rigid registration. FeT consists of: 1) a structure variability measure S_{var} defined by the trace of a covariance matrix $\mathcal{C}$ for seven features: Position (x, y), absolute values of first and second order image derivatives ($|g_x|, |g_y|, |g_{xx}|, |g_{yy}|$), and intensity difference $|g(\mathbf{x}) - g(\mathbf{x}_k)|$ for each $\mathbf{x}_k$ within the patch $P_{\mathbf{x}}$, and 2) a texture measure S_T based on cross-correlation between the pixels in $P_{\mathbf{x}}$. A patch size size of 5×5 pixels was chosen after our experimental observations. The combined feature transform (FeT) is then designed as a weighted sum of the two components $FeT = S_{var} + 0.5 S_T$. Figure 3 (right) shows example results for FeT for blockface (FeT_{g_1}) and PLI images (FeT_{g_2}). It can be seen that boundaries and inner texture are quite similar for the multi-modal images. To preserve discontinuities of the deformation field, we use Gaussian weights f_σ for the quadratic regularization. We use the energy functional

$$\underset{\mathbf{u},\mathbf{u}^I}{\text{argmin}} \sum_\Omega \underbrace{J_{data}(FeT_{g_1}, FeT_{g_2}, \mathbf{u}^I) + \lambda_I f_\sigma(\|\mathbf{u}\|) \, \|\mathbf{u} - \mathbf{u}^I\|_2^2}_{J_{Intensity}} + \lambda_E \, J_{elastic}(\mathbf{u}) \,,$$

(3)

where FeT_{g_1} and FeT_{g_2} are the feature transforms of the target and source images, respectively. The weighting factors $\lambda_I, \lambda_E > 0$ control the trade-off between the data term J_{data} and the two regularization terms (quadratic and elastic, in our case $\lambda_I = 0.25$, $\lambda_E = 0.25$). Here, quadratic regularization will allow for smoother deformation field while elastic regularization will force for a realistic deformation and avoid any unnatural deformations. $\mathbf{u}^I$ is the deformation field obtained by minimizing the SSD between the feature transforms with a weighted quadratic regularization (i.e. minimization of $J_{Intensity}$) using Levenberg-Marquardt optimization. The final deformation field $\mathbf{u}$ is obtained using an analytic solution based on GEBS.

Figure 4 (left) reveals the result after rigid registration. Visual inspection shows a good alignment, however, misalignments are distinct along the corpus callosum (indicated by blue arrows), the hippocampus (cyan arrows), and along the borders of the cerebral cortex (black arrows). Using the new similarity measure for non-rigid registration, it can be observed in Fig. 4 that the misalignments in various regions have been tackled (see Fig. 4, middle and right).

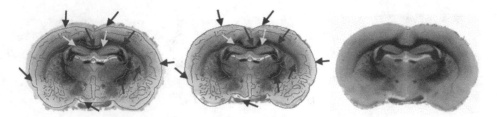

Fig. 4. Registration of high-resolution 3D-PLI data. Left: Rigid registration, middle: non-rigid registration (edges of blockface overlaid with high-resolution 3D-PLI image), and right: Color-overlay image with blockface (green) and registered high-resolution3D-PLI image (red). (Color figure online)

2.2   Registration of Ultra-High Resolution 3D-PLI Data

Due to the large difference in spatial resolution between the blockface images and the ultra-high resolution images (factor of about 12) and arbitrary rotations we perform registration using a scale-space method for feature detection and matching. A Gaussian scale-space and a Hessian measure are used to detect features in the registered high-resolution and the ultra-high resolution (retardation map) images with subsequent feature matching based on FLANN [2]. Then, a similarity transformation (rigid and isotropic scaling) is computed using the matched features and a least squares approach. Unlike in [2], we use a fast bilateral

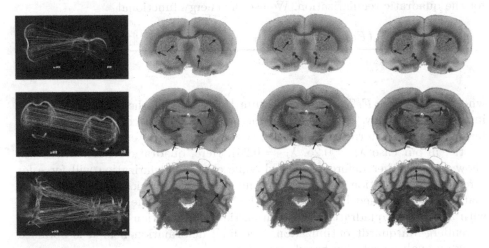

Fig. 5. Registration of ultra-high resolution (u-HR) images to the registered high-resolution (HR) images. First column: Matching of detected salient features (retardation maps), second column: Alignment after similarity transformation, third column: Alignment after non-rigid registration (transmittance maps), and fourth column: Highlighted regions with large deformations. Misalignments are indicated by green regions and black arrows. (Color figure online)

filtering technique [12] to cope with the noise in the rat brain data and reduce false detections in feature extraction. In Fig. 5, examples for feature matching results are shown. Subsequently, a non-rigid registration (see Sect. 2.1) is used to cope with local deformations at 15.5 μm resolution (see Fig. 5, third column). Further, visible misalignments in Fig. 5 (third column) are corrected at a resolution of 1.3 μm using the proposed non-rigid registration method (see Eq. (3)) and coarse-to-fine energy minimization (we use 9 pyramid levels).

Table 1. Mean target registration error and standard deviation (TREs±std. dev.) using landmarks (LMs) from an expert for three rat brain sections. Registration of high-resolution (HR) images and ultra-high-resolution (u-HR) images to reference blockface (15.5 μm).

Section	LMs		HR($64\mu m$) ⇒ BF				u-HR($1.3\mu m$) ⇒ BF			
			initial	rigid		non-rigid	initial	rigid	non-rigid	
	HR	u-HR	COM	g	$\tilde{g}$	MI	FeT	COM + scale	MI	FeT
# 105	25	35	708.3	37.7	23.9	**7.0**	7.2	419.7	8.4	13.5	**4.9**
			±93.4	±16.8	±14.7	±6.7	±3.2	±13.4	±3.5	±6.8	±2.9
# 131	22	42	785.8	40.3	22.2	13.9	**6.0**	532.7	25.7	13.3	**10.4**
			±247.5	±10.6	±6.8	±3.5	±3.3	±271.0	±9.9	±5.7	±5.9
# 337	29	61	701.6	25.0	20.9	11.4	**6.7**	453.1	21.6	10.9	**7.9**
			±183.7	± 8.1	±7.3	±11.0	±3.3	±264.6	±7.7	±9.0	±4.5
Mean:	25	46	731.9	34.3	22.3	10.7	**6.6**	475.4	18.6	12.5	**7.7**
			±174.8	±11.8	±9.6	±7.0	±3.3	±173.0	±7.0	±7.2	±4.4

3 Experimental Results

We have evaluated the proposed method for the registration and 3D reconstruction of high- and ultra-high-resolution data of the rat brain (64 μm and 1.3 μm). Ground truth correspondences for three sections were determined manually by an expert (on average 25 and 46 landmarks for high- and ultra-high resolution sections, respectively). Table 1 shows the average target registration error (TRE). It can be seen that our proposed non-rigid registration method using the feature transform FeT yielded an overall improvement of about 4.1 pixels and 4.8 pixels compared to a previous non-rigid registration approach using mutual information [6] for high- and ultra-high resolution 3D-PLI data, respectively. Notably, our non-rigid registration method can deal with large deformations which is evident from the large overall improvements of 15.7 pixels and 10.9 pixels compared to rigid registration using CoT ($\tilde{g}$) for high-resolution and ultra-high resolution data, respectively. Also, Fig. 6 shows that our non-rigid method is able to cope with highly non-linear deformations present in our full rat brain data (in range from 0–132 pixels in magnitude).

Figure 7 (left, middle) shows 3D visualizations of registration results as a reconstructed 3D volume of 278 high-resolution image sections (transmittance

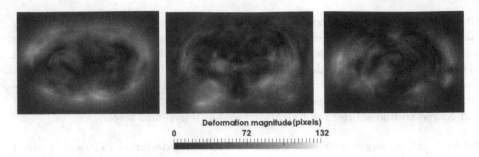

Deformation magnitude (pixels)

0   72   132

Fig. 6. Deformation magnitudes after correction using our non-rigid registration approach on high-resolution PLI data for 3 different sections (#131, #210, #283).

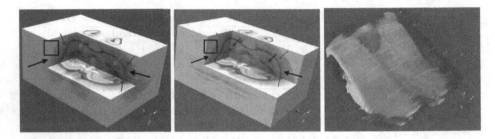

Fig. 7. 3D reconstruction. Left: Rigid registration, middle: non-rigid registration, and right: rendered 3D volume at 1.3 μm resolution (scaled for visualization). (Color figure online)

maps, 15.5 μm × 15.5 μm × 16.7 mm). After rigid registration, misalignments are visible at locations indicated by arrows (black: Tissue boundary, blue: Corpus callossum and red: Caudate putamen) and a square in Fig. 7 (left). However, after non-rigid registration a coherent alignment can be observed (see Fig. 7, middle). A rendered 3D reconstructed volume of ultra-high resolution is shown in Fig. 7 (right) where the smooth green regions indicate coherent alignment of corpus callossum (retardation maps, 1.3 μm × 1.3 μm × 16.7 mm).

All the implementations are in C++ and we have used optimized C++ libraries for computing trace of covariance matrix to speed-up the FeT-based SSD metric minimization in our non-rigid approach. Additionally, entire framework, that is from high-resolution to ultra-high resolution reconstruction, is built as a parallel processing pipeline and optimized for speed-up in complete 3D reconstruction.

4   Conclusion

We have introduced a new multi-scale and multi-modal registration method for 3D reconstruction of both high-resolution and ultra-high resolution 3D-PLI histological images of a rat brain. The method comprises a novel feature transform-based similarity metric integrated in a physically-based non-rigid registration approach as well as a correlation transform-based similarity measure for robust rigid registration. Quantitative evaluations showed that our method improves the result compared to a previous multi-modal non-rigid registration approach and leads to a coherent 3D reconstruction.

Acknowledgments. This project was funded by the Helmholtz Association through the Helmholtz Portfolio theme "Supercomputing and Modeling for the Human Brain" and by the European Union through the Horizon 2020 Research and Innovation Programme under Grant Agreement No. 7202070 (Human Brain Project SGA1).

References

1. Ali, S., et al.: Elastic registration of high-resolution 3D PLI data of the human brain. In: Proceedings of 14th IEEE International Symposium on Biomedical Imaging (ISBI), Melbourne, Australia, 18–21 April, pp. 1151–1155 (2017)
2. Ali, S., Rohr, K., Axer, M., Amunts, K., Eils, R., Wörz, S.: Registration of ultra-high resolution 3D PLI data of human brain sections to their corresponding high-resolution counterpart. In: Proceedings of 14th IEEE International Symposium on Biomedical Imaging (ISBI), Melbourne, Australia, 18–21 April, pp. 415–419 (2017)
3. Ali, S., Wörz, S., Amunts, K., Eils, R., Axer, M., Rohr, K.: Rigid and non-rigid registration of polarized light imaging data for 3D reconstruction of the temporal lobe of the human brain at micrometer resolution. NeuroImage **181**, 235–251 (2018)
4. Arsigny, V., Pennec, X., Ayache, N.: Polyrigid and polyaffine transformations: a new class of diffeomorphisms for locally rigid or affine registration. In: Ellis, R.E., Peters, T.M. (eds.) MICCAI 2003. LNCS, vol. 2879, pp. 829–837. Springer, Heidelberg (2003). https://doi.org/10.1007/978-3-540-39903-2_101
5. Axer, M., et al.: A novel approach to the human connectome: ultra-high resolution mapping of fiber tracts in the brain. NeuroImage **54**(2), 1091–1101 (2011)
6. Biesdorf, A., Wörz, S., Kaiser, H.-J., Stippich, C., Rohr, K.: Hybrid spline-based multimodal registration using local measures for joint entropy and mutual information. In: Yang, G.-Z., Hawkes, D., Rueckert, D., Noble, A., Taylor, C. (eds.) MICCAI 2009. LNCS, vol. 5761, pp. 607–615. Springer, Heidelberg (2009). https://doi.org/10.1007/978-3-642-04268-3_75
7. Chou, P., Pagano, N.: Elasticity - Tensor, Dyadic, and Engineering Approaches. Dover Publications Inc., Mineola (1992)
8. Kohlrausch, J., Rohr, K., Stiehl, H.: A new class of elastic body splines for nonrigid registration of medical images. J. Math. Imaging Vis. **23**(3), 253–280 (2005)
9. Lebenberg, J., et al.: Validation of MRI-based 3D digital atlas registration with histological and autoradiographic volumes: an anatomofunctional transgenic mouse brain imaging study. NeuroImage **51**, 1037–1046 (2010)
10. Majka, P., Wójcik, D.: Possum-a framework for three-dimensional reconstruction of brain images from serial sections. Neuroinformatics **14**(3), 265–278 (2016)
11. Ourselin, S., Roche, A., Subsol, G., Pennec, X., Ayache, N.: Reconstructing a 3D structure from serial histological sections. Image Vis. Comput. **19**(1), 25–31 (2001)
12. Paris, S., Durand, F.: A fast approximation of the bilateral filter using a signal processing approach. Internat. J. Comput. Vis. **81**(1), 24–52 (2009)
13. Schubert, N.: 3D reconstructed cyto-, muscarinic M2 receptor, and fiber architecture of the rat brain registered to the waxholm space atlas. Front. in Neuroanat. **10**, 51 (2016)

14. Thévenaz, P., Ruttimann, U., Unser, M.: A pyramid approach to subpixel registration based on intensity. IEEE Trans. Image Process. **7**(1), 27–41 (1998)
15. Wörz, S., Rohr, K.: Spline-based hybrid image registration using landmark and intensity information based on matrix-valued non-radial basis functions. Int. J. Comput. Vis. **106**(1), 76–92 (2014)

FOD-Based Registration
for Susceptibility Distortion Correction
in Connectome Imaging

Yuchuan Qiao, Wei Sun, and Yonggang Shi[✉]

USC Stevens Neuroimaging and Informatics Institute,
Keck School of Medicine of USC, University of Southern California,
Los Angeles, USA
Yonggang.Shi@loni.usc.edu

Abstract. Multi-shell, high resolution diffusion MRI (dMRI) data from the Human Connectome Project (HCP) provides an unprecedented opportunity for the *in vivo* mapping of human brain pathways. It was recently noted, however, that significant distortions remain present in the data of most subjects preprocessed by the HCP-Pipeline, which have been widely distributed and used extensively in connectomics research. Fundamentally this is caused by the reliance of the HCP tools on the B0 images for registering data from different phase encodings (PEs). In this work, we develop an improved framework to remove the residual distortion in data generated by the HCP-Pipeline. Our method is based on more advanced registration of fiber orientation distribution (FOD) images, which represent information of dMRI scans from all gradient directions and thus provide more reliable contrast to align data from different PEs. In our experiments, we focus on the brainstem area and compare our method with the preprocessing steps in the HCP-Pipeline. We show that our method can provide much improved distortion correction and generate FOD images with more faithful representation of brain pathways.

1   Introduction

With the advance of multiband and several other MRI techniques, the Human Connectome Project (HCP) [1] has developed cutting-edge connectome imaging protocols to acquire high-resolution, multi-shell diffusion MRI (dMRI) that has enabled the *in vivo* investigation of brain pathways with unprecedented details. Among the various technical advances in the HCP protocol, one notable choice is the acquisition of dMRI data from two phase encodings (PEs) for the correction of susceptibility-induced distortion. However, it was noted recently that majority of the preprocessed dMRI data from HCP [2] still contains significant distortions in regions such as the brainstem [3]. Given the critical role that HCP data

This work was in part supported by the National Institute of Health (NIH) under Grant R01EB022744, R01AG056573, U01EY025864, P41EB015922.

G. Wu et al. (Eds.): CNI 2018, LNCS 11083, pp. 11–19, 2018.
https://doi.org/10.1007/978-3-030-00755-3_2

continues to play in brain imaging research, it is important to raise awareness of this critical issue in the neuroimaging community and investigate solutions to this fundamental problem.

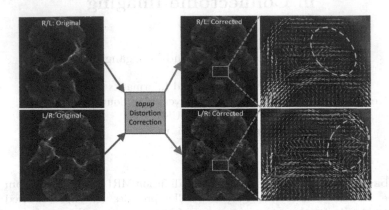

Fig. 1. An illustration of the problem in susceptibility distortion correction in HCP-Pipeline. Inputs to the distortion correction based on the *topup* tool from FSL are the B0 images from the R/L and L/R PE. After that, the corrected B0 images appear "undistorted". For the ROIs (yellow box) of the corrected data from each PE, we computed the FODs and plotted them in the right column. The residual distortions are clearly visible as highlighted in the dashed ellipsoids. (Color figure online)

The susceptibility of the magnetic field to the tissue/air boundary around the brainstem results in large distortions of the dMRI data along the phase encoding (PE) direction [4]. As shown in Fig. 1, the distortion can be either stretching or compression of the brainstem tissue. The susceptibility-induced distortion not only causes geometric distortion, but more importantly the loss of information due to piling up of signal intensities in regions with severe compression. The state-of-the-art solution is to collect data from opposite PE directions and estimate the distortion field from these two copies of data with the same set of gradient directions [5–7], which is the approach HCP and various connectome imaging projects adopt. Using data from the two PEs, the HCP has developed the HCP-Pipeline to correct the distortions and merge them to generate the preprocessed dMRI data for connectivity research [2], which has been widely distributed to the research community. The susceptibility distortion correction in the HCP-Pipeline is mainly based on the topup tool from the FSL, which registers the B0 images from both PEs and estimates the deformation fields for the correction. From the fiber orientation distributions (FODs) [8] plotted in Fig. 1, we can clearly see the brainstem anatomy is severely misaligned even though the "corrected" B0 images appear undistorted. This is because the B0 images do not have enough contrast to resolve the complex brainstem anatomy. Merging such misaligned data, if we continue to run the preprocessing steps in the HCP-Pipeline, will clearly introduce severe artifacts for connectivity modeling. The critical issue is that this is not a rare problem, but prevalent in most preprocessed data [3]. In the benign cases, we might expect distorted pathways. In more serious situations, false pathways can be introduced as we will demonstrate in our experiments. It is thus imperative to tackle this problem and provide improved correction for the highly valuable HCP dataset.

To overcome limitations in current methods for susceptibility distortion correction, we will develop a novel method based on registering the FODs computed from data acquired in each PE. Compared with B0 images which existing methods rely on for distortion correction, FODs represent information from all gradient directions and encode more detailed contrast to allow better alignment of brainstem anatomy. Our method first estimates the FODs for the multi-shell imaging data in each PE. A variational optimization approach will then be developed to minimize the mismatch of FODs from the two PEs while requiring the distortion fields to be opposite transformations as much as possible, which stems from the physical models of susceptibility distortions. After that, we solve a regularized inverse problem to merge these corrected data for connectivity analysis. Compared with previous works based on registering B0 images, there are two main advantages in our FOD-based approach. Because FODs are computed from the ratio between images of each gradient direction to the B0 image, and images from all gradient directions experience the same susceptibility distortion in one acquisition, the Jacobian modulation to image intensities are not needed in our formulation. The second advantage is that there no need of incorporating the rotation operator during the deformation process because the distortion is one-dimensional along the PEs, which leads to a much simplified FOD registration framework. In our experiments, we compare our method with the HCP-Pipeline and demonstrate that much improved distortion correction can be achieved on HCP data.

2   Methods

An overview of the proposed novel method for susceptibility distortion correction is shown in Fig. 2. The inputs to the processing pipeline are the dMRI data from two opposite PEs of the same subject. For data from the HCP, the two PEs are R/L and L/R. In most other connectome imaging studies, the two PEs are typically the A/P and P/A directions. In the first part of the workflow (purple dashed box), the processing steps of the HCP-Pipeline are first applied including the topup and eddy tools from the FSL. In the default HCP-Pipeline, the "corrected" data from both PEs will then be merged to form the output dMRI data. The novel processing pipeline developed in this work is shown in the cyan box in Fig. 2. To remove the residual distortion in the data processed by HCP-Pipeline, we compute the FODs from the multi-shell dMRI data of both PEs to better characterize the differences in these images. The deformation fields from FOD-based registration are then extracted to correct for distortions and merging the data from the two PEs into a final dMRI dataset for connectivity analysis.

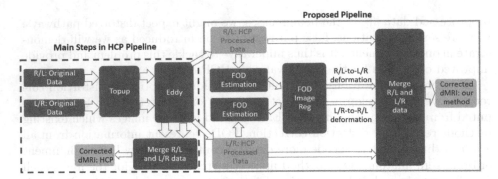

Fig. 2. The proposed framework for susceptibility distortion correction for connectome dMRI data. (Color figure online)

2.1 FOD Image Registration

For the multi-shell dMRI data from each PE, we compute the FODs using the multi-compartment model in [8]. At each voxel, the FOD is represented as the coefficients to the spherical harmonics (SPHARMs). For the registration of the FOD images from two PEs, we use the open source software package `elastix` [9] and solve an image registration optimization problem as follows:

$$\widehat{\boldsymbol{\mu}} = \arg\min_{\boldsymbol{\mu}} \mathcal{C}(I_F, I_M \circ \boldsymbol{T}(\boldsymbol{x}, \boldsymbol{\mu})), \tag{1}$$

where I_F and I_M are the fixed and moving image, respectively, $\boldsymbol{x}$ is an image voxel location and $\boldsymbol{T}(\boldsymbol{x}, \boldsymbol{\mu})$ is a coordinate transformation parameterized by $\boldsymbol{\mu}$. In our work, both the fixed and moving images are 4-D images with the 4-th dimension denoting the SPHARM basis functions used for FOD representation. Practically we took the first N SPHARM coefficient images of each PE to form the 4-D vector images for registration. The image mask was generated from the HCP-Pipeline [2]. The mutual information is used as a cost function, and B-spline transformation model is chosen for the modeling of the deformation. An accelerated version of adaptive stochastic gradient descent [10] was used for iterative optimization of Eq. (1). A multi-resolution strategy was used to tackle the local minimum and accelerate the registration procedure. The images were smoothed using a half-reduced Gaussian filter with standard deviations from 32 to 0.5 mm. For each iteration, 10000 image voxels were randomly sampled from the fixed image, and 1500 iterations were used for each resolution.

Following previous works on modeling the susceptibility distortion [5–7], we constrained the deformation only along the phase encoding direction during the optimization of the registration. For the HCP data, we denote the FOD from R/L PE as the fixed image, and FOD from L/R PE as the moving image. Using the above registration process, we compute the forward deformation $\mathbf{D}$ from R/L to L/R and the inverse deformation $\text{inv}(\mathbf{D})$ from L/R to R/L. For data from opposite PEs, the susceptibility distortion are symmetric in opposite PE direction. For the preprocessed data from HCP-Pipeline, we assume that the residual distortions from the two PEs are still symmetric in each PE direction since the symmetry was modeled in the *topup* tool [5]. The underlying true, un-distorted image should thus be at the middle point between the fixed and moving image. Half-way deformations $\mathbf{D}/2$ and $inv(\mathbf{D})/2$ will thus be used to merge the data from two PEs and reconstruct the true dMRI image next.

2.2 Merging Data from Different PEs

Using the deformations from FOD-based registration, we will solve an inverse problem to reconstruct the underlying true image I from two distorted images, denoted as Y_+ and Y_-, from two different PEs. Because we assume the I is at the middle point between the fixed Y_+ and moving image Y_-, we can use the half-way deformations $\mathbf{D}/2$ and $inv(\mathbf{D})/2$ and formulate a forward model as:

$$\begin{bmatrix} Y_+ \\ Y_- \end{bmatrix} = \begin{bmatrix} K_+ \\ K_- \end{bmatrix} I + n, \tag{2}$$

where K_+ and K_- are sparse matrix representations of the deformations $\mathbf{D}/2$ and $inv(\mathbf{D})/2$ that relates the true image I to the images from two opposite PEs, and n denotes noise. Using regularized least squares, we can obtain the solution to the inverse problem as

$$I = \left(\begin{bmatrix} K_+^T & K_-^T \end{bmatrix} \begin{bmatrix} K_+ \\ K_- \end{bmatrix} + \lambda R \right)^{-1} \begin{bmatrix} K_+^T & K_-^T \end{bmatrix} \begin{bmatrix} Y_+ \\ Y_- \end{bmatrix}, \tag{3}$$

where R is a smoothness regularization and λ is a non-negative parameter controlling the relative weight of the regularization term. This reconstruction process is applied to data from each gradient direction and form the final output data from our method, which removes the residual distortions and produces high quality data for connectivity analysis.

3   Experimental Results

In this section, we will apply our method to the connectome imaging data from HCP and compare the performance with HCP-Pipeline. In particular, we will focus on the brainstem area, which is a critical region for brain imaging research but usually experiences high distortion. More recently, the brainstem area is considered the earliest site of *tau* pathology in the Braak staging of Alzheimer's disease (AD) [11], thus making it highly significant to obtain accurate dMRI imaging data in this challenging area of human brain.

For each HCP subject, we use two raw dMRI scans in the R/L and L/R PEs from the 900-subject release of HCP. Each dMRI scan acquires data from 97 gradient directions distributed on three shells with b-values 1000, 2000, and 3000 s/mm^2 at an isotropic spatial resolution of 1.25 mm. For each subject, we will compare the FOD fields with known anatomy in the brainstem to evaluate reconstruction quality. For all experiments, the FODs are calculated with the

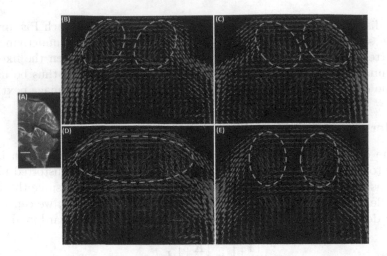

Fig. 3. Results from a brainstem ROI at the level of the midpons of subject 100307. (A) Yellow bar indicates the location of the ROI on the mid-sagittal slice. (B) and (C) show the FODs computed with data from the R/L and L/R PE after preprocessed by HCP-Pipeline, respectively. (D) show the FODs computed with the corrected data from HCP-Pipeline after merging data from R/L and L/R PEs. (E) show the FODs computed with the merged data from our method. (Color figure online)

maximum SPHARM order of 12. The regularization parameter in the merging process is chosen as 0.5. We have conducted our experiments on 10 randomly selected HCP subjects. Overall we observe clear improvements in all subjects after we applied our processing workflows to data generated by HCP-Pipeline. This is even true for subjects that passed the quality control of [3]. Next we show results from two representative subjects to demonstrate the improved distortion corrections achieved by our method.

We first show that residual distortions can still exist in data considered to be high quality. In Fig. 3, we plotted the FODs computed from data generated at different stages of the workflow shown in Fig. 2 for subject 100307, which has passed the quality control in [3] based on visual examination of tract density images (TDI). In both Fig. 3(B) and (C), the dashed ellipsoids highlight the FODs for the left and right cortico-spinal-tract (CST). Clearly they are oriented differently in Fig. 3(B) and (C), which indicates the presence of residual distortions. If we merge these two datasets, as done typically by HCP-Pipeline, and computed the FODs as shown in Fig. 3(D), the FODs corresponding to the left and right CST become mingled together, which is not anatomically correct since the CSTs only decussate at the more inferior medulla area of the brainstem. In the FODs computed from the data generated by our method shown in Fig. 3(E), we highlight that the FODs for the left and right CST in dashed ellipsoids. This shows that our method not only maintains the separation of the two CSTs in midpons, but also corrects the distortions that existed in the data generated by HCP-Pipeline.

In addition to the relatively benign distortions such as in subject 100307, we also observe more critical cases that false fiber trajectories can be caused in the merged data if the distortions are not removed. In Fig. 4, we show the results from subject 136833 at an ROI in the midbrain. In Fig. 4(B) and (C), we show the FODs computed from the data of each PE after they went through the distortion correction steps of HCP-Pipeline. The two dashed ellipsoids highlight the trajectories of the fiber tracts of the cerebellar peduncles on the left and right hemispheres, but they are obviously distorted toward the right and left side, respectively. If we simply merge the data after the distortion correction in HCP-Pipeline, the resulting FODs shown in Fig. 4(D) suggest there are two possible fiber trajectories in each hemisphere as highlighted by the dashed ellipsoid. This is due to the merge of the misaligned data from two PEs and clearly not anatomically meaningful. After we apply the distortion correction by our method, we can see the FOD trajectories of these two fiber pathways become much better aligned and maintains the same number of pathways as in the data from both PEs.

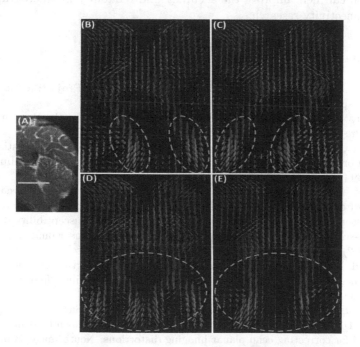

Fig. 4. Results from a brainstem ROI at the level of the midbrain of subject 136833. (A) Yellow bar indicates the location of the ROI on the mid-sagittal slice. (B) and (C) show the FODs computed with data from the R/L and L/R PE after preprocessed by HCP-Pipeline, respectively. (D) show the FODs computed with the corrected data from HCP-Pipeline after merging data from R/L and L/R PEs. (E) show the FODs computed with the merged data from our method. (Color figure online)

4 Conclusions and Discussion

There are two main goals of this work. The first is to raise awareness of the residual distortions in the connectome imaging data from HCP and other imaging projects even after they are processed by the HCP-Pipeline. As we showed in our results, these distortions can lead to detrimental effects for modeling the connectivity of brain pathways. The second goal of this work is the development of a novel framework for the correction of these residual distortions in connectome imaging data. Our method is based on the registration of FOD images from both phase encodings, which provides more anatomically relevant contrast than B0 images used in current tools in the HCP-Pipeline. We demonstrated that our method can remove residual distortions and produce anatomically more valid FODs for brainstem regions. Besides the brainstem area, connectome imaging data have distortions throughout the brain, even for deep brain areas around the ventricles. As our future work, it is thus important to conduct a more extensive examination of the residual distortions in other brain regions, and study how our method can help improve the accuracy and reliability in tractography and related connectivity research.

References

1. Essen, D.V., Ugurbil, K., et al.: The Human Connectome Project: a data acquisition perspective. NeuroImage **62**(4), 2222–2231 (2012)
2. Glasser, M.F., et al.: The minimal preprocessing pipelines for the Human Connectome Project. NeuroImage **80**, 105–124 (2013)
3. Tang, Y., Sun, W., Toga, A.W., Ringman, J.M., Shi, Y.: A probabilistic atlas of human brainstem pathways based on connectome imaging data. NeuroImage **169**, 227–239 (2018)
4. Jezzard, P., Balaban, R.S.: Correction for geometric distortion in echo planar images from B0 field variations. Magn. Reson. Med. **34**(1), 65–73 (1995)
5. Andersson, J.L., Skare, S., Ashburner, J.: How to correct susceptibility distortions in spin-echo echo-planar images: application to diffusion tensor imaging. NeuroImage **20**(2), 870–888 (2003)
6. Holland, D., Kuperman, J.M., Dale, A.M.: Efficient correction of inhomogeneous static magnetic field-induced distortion in echo planar imaging. NeuroImage **50**(1), 175–183 (2010)
7. Irfanoglu, M.O., Modi, P., Nayak, A., Hutchinson, E.B., Sarlls, J., Pierpaoli, C.: DR-BUDDI (Diffeomorphic Registration for Blip-Up blip-Down Diffusion Imaging) method for correcting echo planar imaging distortions. NeuroImage **106**, 284–299 (2015)
8. Tran, G., Shi, Y.: Fiber orientation and compartment parameter estimation from multi-shell diffusion imaging. IEEE Trans. Med. Imag. **34**(11), 2320–2332 (2015)
9. Klein, S., Staring, M., Murphy, K., Viergever, M.A., Pluim, J.P.: Elastix: a toolbox for intensity-based medical image registration. IEEE Trans. Med. Imaging **29**(1), 196–205 (2010)

10. Qiao, Y., van Lew, B., Lelieveldt, B.P., Staring, M.: Fast automatic step size estimation for gradient descent optimization of image registration. IEEE Trans. Med. Imaging **35**(2), 391–403 (2016)
11. Braak, H., Thal, D.R., Ghebremedhin, E., Del Tredici, K.: Stages of the pathologic process in Alzheimer disease: age categories from 1 to 100 years. J. Neuropathol. Exp. Neurol. **70**(11), 960–969 (2011)

GIFE: Efficient and Robust Group-Wise Isometric Fiber Embedding

Junyan Wang[✉] and Yonggang Shi

Laboratory of Neuro Imaging (LONI), USC Stevens Neuroimaging
and Informatics Institute, Keck School of Medicine of University
of Southern California, Los Angeles, CA 90033, USA
junyan.wang@loni.usc.edu

Abstract. Tractography is a prevalent technique for in vivo imaging of
the white matter fibers (a.k.a. the tractograms), but it is also known to be
error-prone. We previously propose the Group-wise Tractogram Analysis
(GiTA) framework for identifying anatomically valid fibers across subjects according to cross-subject consistency. However, the original framework is based on computationally expensive brute-force KNN search. In
this work, we propose a more general and efficient extension of GiTA. Our
main idea is to find the finite dimensional vector-space representation of
the fiber tracts of varied lengths across different subjects, and we call
it the group-wise isometric fiber embedding (GIFE). This novel GIFE
framework enables the application of the powerful and efficient vector
space data analysis methods, such as the k-d tree KNN search, to GiTA.
However, the conventional isometric embedding frameworks are not suitable for GIFE due to the massive fiber tracts and the registration errors
in the original GiTA framework. To address these issues, we propose a
novel method called multidimensional extrapolating (MDE) to achieve
GIFE. In our experiment, simulation results show quantitatively that our
method outperforms the other methods in terms of computational effi-
ciency/tractability and robustness to errors in distance measurements
for real fiber embedding. In addition, real experiment for group-wise
optic radiation bundle reconstruction also shows clear improvement in
anatomical validity of the results from our MDE method for 47 different
subjects from the Human Connectome Project, compared to the results
of other fiber embedding methods.

1 Introduction

Tractography computes white matter fiber tracts from diffusion MRI and it is
a prevalent technique for in-vivo measurement of anatomical connectivity of the
brain. However, this technique is also known to be error-prone [1].

Fiber filtering methods have been proposed previously to remove redundant
and anatomically invalid fibers based on data fidelity [2], geometrical sound-
ness [3] and anatomical knowledge [4]. Yet, these methods did not explicitly

This work was in part supported by the National Institute of Health (NIH) under
Grant R01EB022744, R01AG056573, U01EY025864, P41EB015922.

© Springer Nature Switzerland AG 2018
G. Wu et al. (Eds.): CNI 2018, LNCS 11083, pp. 20–28, 2018.
https://doi.org/10.1007/978-3-030-00755-3_3

address the inter-subject consistency and the results are not guaranteed to be consistent across subjects, or reproducible. *We hypothesize that the errors in the fiber tracts computed with tractography are random and they are not anatomically consistent across different subjects.* Accordingly, we recently proposed a data-driven framework called Group-wise Tractogram Analysis (GiTA) which identifies the anatomically valid tracts as the common tracts among different subjects [5]. The idea is to measure the *commonness* of each fiber tract with respect to all different subjects and identify those that are *common* among most of the subjects. A major limitation of GiTA is that this framework is computationally very expensive, as it requires comparing all tracts across all subjects. In addition, the efficient and powerful data analysis methods based on vector-space data representation, such as the k-d tree KNN search, are not applicable due to the unequaled lengths of the fiber tracts. It is also invalid to resampled the fiber curves to the same number of points for computing the Euclidean distance since this generally doesn't reflect the intrinsic geometrical distance between the fiber curves. Besides, the GiTA framework also suffers from inter-subject misalignment due to registration error.

To address the aforementioned issues, we propose a novel group-wise isometric fiber embedding (GIFE) framework. The GIFE framework tries to find the finite dimensional embedding of the fiber tracts of any target subject, given the pre-computed embedding of the fiber tracts for the reference subject. This GIFE framework is naturally parallelizable and it does not require computing the full pairwise distance matrices across all pairs of subjects but only the distances between the fibers of the target subjects and the reference subject. Furthermore, we also handle the inter-subject misalignment in the embedding and derive a novel method called multidimensional extrapolating (MDE), as a tribute to the original multidimensional scaling (MDS) framework, to achieve robust and efficient GIFE. MDE can be viewed as a novel variant of the multidimensional unfolding (MDU) framework [6,7]. Unlike the conventional MDU, MDE allows the embedding of the reference set to be fixed. Besides, MDE deals specifically with the errors in distance measurements such as registration error.

Previously, fiber embedding based on tract affinity has been applied to fiber bundle segmentation [8] without preserving the distance in the embedding space. The embedding based on MDS has been applied to fiber visualization for individual subjects [9]. However, it is not scalable to GiTA.

2   GIFE: Group-Wise Isometric Fiber Embedding

We adopt the principle of MDS for GIFE because it preserves pairwise distances. In addition, we address the scalability by using only a small amount of tract distances. Moreover, since the inter-subject distances are often inaccurate in our problem, rather than solving the embedding from the inter-subject distances directly, we transform the embedding found using the intra-subject tract distances to fit to the geometry defined by the inter-subject distances and the resultant embedding is robust to the errors in inter-subject distances.

2.1   The Classical MDS

Our idea is based on the classical MDS. The basic formulation of the classical MDS can be written as follows [6]:

$$\mathbf{B}^{n \times n} \approx \mathbf{Z}_p \mathbf{Z}_p^T \tag{1}$$

where $\mathbf{B}^{n \times n} = -\frac{1}{2} P^{n \times n} \mathbf{D}^{(2)n \times n} P^{n \times n}$, $\mathbf{Z}_p$ is the p-dimensional embedding of the fiber tracts, and it denotes the first p columns of the matrix $\mathbf{Z}$, P is known as the centering matrix defined as $P_{ij} = 1 - \frac{1}{n}, \forall i = j$ and $P_{ij} = -\frac{1}{n}, \forall i \neq j$, $\mathbf{D}^{(2)}$ is the input squared distance matrix.

The solution of Eq. (1) can be obtained by using the following steps:

$$(a) \ \mathbf{B} = P\left[-\frac{1}{2}\mathbf{D}^{(2)}\right] P, \ (b) \ \mathbf{E}\mathbf{\Lambda}\mathbf{E}^T = \mathtt{svd}(\mathbf{B}), \ (c) \ \mathbf{Z}_p = \mathbf{E}_p \mathbf{\Lambda}_p^{\frac{1}{2}} \tag{2}$$

where $\mathbf{Z}_p$ and $\mathbf{E}_p$ are the first p columns of $\mathbf{Z}$ and $\mathbf{E}$, $\mathbf{\Lambda}_p$ is the top-left $p \times p$ block matrix of $\mathbf{\Lambda}$.

To simplify the derivations and implementation, we make the following assumption.

Assumption 1. *Let* $-\frac{1}{2}\mathbf{D}^{(2)} = \mathbf{V}\mathbf{A}\mathbf{V}^T$, *where* $\mathbf{V}$ *and* $\mathbf{A}$ *are the eigenvector and eigenvalue matrices of* $-\frac{1}{2}\mathbf{D}^{(2)}$. *We assume* $\mathbf{V}_p = P\mathbf{V}_p$.

In fact, we can always impose $\mathbf{V} = P\mathbf{V}$ as a constraint in the classical MDS factorization framework. We omit this step in this work since we observe that this complication unnecessary as the assumption is often valid in our problem.

Based on this assumption, we have:

$$\mathbf{B} = P\left[-\frac{1}{2}\mathbf{D}^{(2)}\right] P = -\frac{1}{2}\mathbf{D}^{(2)} = \mathbf{B}^{\#} \tag{3}$$

which is the simplified Gram matrix that we adopt in the rest of this work.

2.2   The Classical Multidimensional Extrapolating (cMDE)

In our problem, computing, storing and factorizing the massive pairwise distances between all fiber tracts for all pairs of subjects are very costly in general. Alternatively, we propose to consider one of the fiber bundles as a reference. Then, for all other bundles, we propose to *estimate* their embedding based on the precomputed embedding of the reference bundle. We call this problem the *GIFE problem*, and we develop a novel method called multidimensional extrapolating (MDE) to solve it. Since we follow the idea of classical MDS so we call our method the classical MDE, or cMDE. There are three variants of the MDE method: inter-set MDE, intra-set MDE and cMDE. The inter-set MDE is computed using only the distances of fibers from different subjects, the intra-set MDE is computed using only the distances from the same subjects, and the cMDE is a combination of these two.

In cMDE, we want to find the embedding $\mathbf{Y}_p$ of the target fibers such that:

$$\begin{bmatrix} \mathbf{X}_p \\ \mathbf{Y}_p \end{bmatrix} [\mathbf{X}_p^T, \mathbf{Y}_p^T] \approx \begin{bmatrix} \mathbf{B}_{XX}^{\#}, \mathbf{B}_{XY}^{\#} \\ \mathbf{B}_{YX}^{\#}, \mathbf{B}_{YY}^{\#} \end{bmatrix} = \mathbf{B}^{\#} = -\mathbf{D}^{(2)} = \begin{bmatrix} -\mathbf{D}_{XX}^{(2)}, -\mathbf{D}_{XY}^{(2)} \\ -\mathbf{D}_{YX}^{(2)}, -\mathbf{D}_{YY}^{(2)} \end{bmatrix} \quad (4)$$

where the reference embedding $\mathbf{X}_p$ of the reference fibers and the simplified Gram matrix $\mathbf{B}^{\#}$ of the reference and target fibers are given.

Inter-set MDE: Embedding with Inter-set Distance. In the ideal case where the registration process in the GiTA is perfect, and the inter-subject distance measure $\mathbf{D}_{XY}$ or $\mathbf{D}_{YX}$ is ideal, we can simply use the following relation to estimate $\mathbf{Y}_p$.

$$\mathbf{B}_{YX}^{\#} \approx \mathbf{Y}_p(\mathbf{X}_p)^T \Rightarrow \widehat{\mathbf{Y}}_p = \mathbf{B}_{YX}^{\#}\mathbf{X}_p\mathbf{\Lambda}_{XX}^{-1} \quad (5)$$

where $\mathbf{X}$ and $\mathbf{\Lambda}_{XX}$ are the eigenvector and eigenvalue matrices of $\mathbf{B}_{XX}^{\#}$, and this formulation is closely related to the *Nyström* approximation [10]. This method might overfit Y to the inter-set distances despite the errors therein.

Intra-set MDE: Embedding with Intra-set Distance. Since $\mathbf{D}_{YY}$ is also known, we can obtain their eigen-decomposition.

$$\mathbf{B}_{YY}^{\#} \approx \mathbf{E}_p^Y \mathbf{\Lambda}_{YY}(\mathbf{E}_p^Y)^T = \overline{\mathbf{Y}}_p\overline{\mathbf{Y}}_p^T \quad (6)$$

The solution found by the above decomposition maybe more reliable than the one found by Eq. (5), since we only use the intra-subject distance measure and no mis-alignment error is present. However, this also does not give us $\mathbf{Y}_p$ directly since

$$\overline{\mathbf{Y}}_p\overline{\mathbf{Y}}_p^T = \overline{\mathbf{Y}}_p RR^T \overline{\mathbf{Y}}_p^T \quad (7)$$

where $RR^T = I^{p \times p}$, meaning that the solution remains valid up to any arbitrary orthogonal transformation.

CMDE: Finding the Optimal Orthogonal Transformation. We propose to estimate the optimal R which transforms the solution of the intra-set MDE $\overline{\mathbf{Y}}$ toward the solution of the inter-set MDE $\widehat{\mathbf{Y}}$, such that the geometry of the embedding defined by $\mathbf{B}_{YY}^{\#}$ is preserved when we try to align the embedding of Y with the inter-set distances. And we propose the following model to solve it:

$$\widetilde{\mathbf{Y}} = \overline{\mathbf{Y}}_p R^*, \; R^* = \arg\min\frac{1}{2}\|R - \overline{\mathbf{Y}}_p^T\widehat{\mathbf{Y}}_p\|^2, \; \text{s.t. } RR^T = I \quad (8)$$

The rationale of this model lies in that $\overline{\mathbf{Y}}_p^T\widehat{\mathbf{Y}}_p$ is actually the least-squares solution of $\min_R \|\overline{\mathbf{Y}}_p R - \widehat{\mathbf{Y}}_p\|^2$ without the orthogonality constraint $RR^T = I$.

The latter formulation is more intuitive. An advantage of Eq. (8) over the latter intuitive formulation is that it admits a closed-form solution [11]

$$R^* = \mathbf{U}\mathbf{V}^T \qquad (9)$$

where $\mathbf{U}\mathbf{D}\mathbf{V}^T = \mathrm{svd}(\overline{\mathbf{Y}}_p^T \hat{Y}_p)$.

3   Experimental Results

3.1   Randomly Sampled Tractograms

In this experiment, we simulate the situation of GIFE problem by subsampling a relatively small real optic radiation bundle reconstructed using the Human Connectome Project (HCP) data. This simulation allows us to quantitatively assess the distance preservability of different methods.

Experiment Configuration. First, we reconstruct the optic radiation fiber bundle for one subject from the HCP data [12,13] using the method described in [14]. This bundle contains a total of 8170 fibers. Note that we used the unfiltered fiber tracts in this experiment, and some spurious tracts are present. We also compute the Hausdorff distance between all pairs of tracts in the bundle. Then, we randomly sample the bundle 10 times without replacement and each subsample contained 10 percent of the original bundle, and the sub-bundles are denoted as $\{T_0, T_2, ..., T_9\}$. Finally, we consider T_0 as the reference bundle and extract its intra-set distance matrix from the full distance matrix, and we extract the intra-set and inter-set distance matrices with reference to T_0 for all other sub-bundles. We mainly compare our method with four different comparable methods for MDU: the weighted least-squares MDS (LS-MDS)[1] [6], the Scaling by MAjorizing a COmplicated Function (SMACOF) algorithm[2] [15] and the Maximum Variance Unfolding (MVU)[3] [7]. Only the inter-subject distances with reference to T_0 and the intra-subject distances are used in this comparison. We also compare our method with the inter-set MDE and the intra-set MDE. Lastly, we compute the cMDS computed with the full distance matrix for reference.

We originally computed cMDS with the full distance matrix using 11 dimensions. However, only 7 of them correspond to positive eigenvalues. According to the MDS theory [6], we should only use positive eigenvalues for cMDS so we fix the dimensionality to be 7 in all methods. In this experiment, we perturbed the inter-set distance D_{T_0,T_i} by $D'_{T_0,T_i} = D_{T_0,T_i} \times (1 + n)$ and $n \sim N(0,0.5)$. Computing the full distance matrix for this bundle took about 11 h on a single CPU core, and computing the subset of distances required by MDE would take only about 4 h.

---

[1] Gradient descent implementation.
[2] http://tosca.cs.technion.ac.il.
[3] http://lvdmaaten.github.io/.

Table 1. Quantitative results for distance preservation in the embedding.

	cMDS	LS-MDS	SMACOF	MVU	inter-set MDE	intra-set MDE	cMDE
$\rho(D, D_{GT})$	0.97	0.48	0.36	0.05	0.72	0.38	0.96
Time (s)	6.38	1562	1445	2723	0.16(/10)	0.6(/10)	0.73(/10)

Results. Some visual results for this simulation experiment are shown in Fig. 1. We observe that our method (cMDE) well approximates the point distribution of cMDS, while the other methods generally fail to preserve the geometrical relationship between the fiber tracts given the partial and imprecise information. From Fig. 2, we can see that the inter-set MDE might be affected by the inter-set misalignment and the intra-set MDE may be oriented arbitrarily while our method optimally restores the point distribution. We are also able to evaluate our method quantitatively by comparing the pairwise distance in the embedding space with the true Hausdorff tract distance. We adopt the signed Pearson correlation coefficient as our distance similarity measure. Note that this measure is invariant to linear scaling which is acceptable in our problem. The results are summarized in Table 1. We observe that our method gives very satisfactory distance preservation in the embedding space. In addition, our method compares significantly favorably to other methods in terms of computational efficiency. The computational cost for calculating the distances are not included in this table.

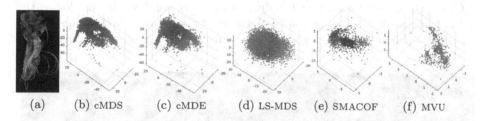

(a)   (b) cMDS   (c) cMDE   (d) LS-MDS   (e) SMACOF   (f) MVU

Fig. 1. 3D visualization of the results of isometric embedding for a randomly sampled visual pathway fiber bundle using different methods. (a) shows the original bundle.

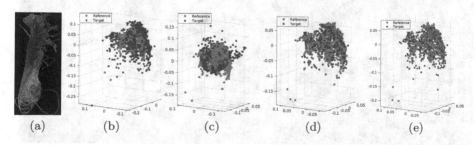

(a)   (b)   (c)   (d)   (e)

Fig. 2. MDE with intermediate results. (a) a target bundle (b) inter-set MDE (c) intra-set MDE (d) cMDE (e) cMDS with full distance matrix.

This experiment is conducted in MATLAB on Linux with Intel(R) Core(TM) i7-6820HQ CPU @ 2.70 GHz and 32 GB memory. All iterative methods terminate at convergence or a maximum of 100 iterations. The computational times shown are all the total times. Since the MDE framework is parallelizable, we put (/10) behind the times to indicate the possibility of further breaking down the computational time by parallelization.

3.2 Common Optic Radiation Fiber Bundle Extraction

Following the GiTA framework, we apply our GIFE framework to extracting common optic radiation fiber bundles [14] using the HCP data for 47 subjects. We use the raw noisy fiber tracts as the input in this experiment. For this task, we pick the fibers common in most of the bundles as the common fiber based on a commonness measure defined based on the Euclidean distance in embedding space or Hausdorff distance in the fiber space. We adopt the k-d tree based KNN search to calculate the Euclidean distances. We use 25 dimensions for the embedding. For results, we expect the common bundle to capture the main anatomical characteristics of the optic radiation bundle and we also expect it to be highly organized to follow retinotopy [4]. The total computation time for calculating all pairwise tract distances was around 60056 hrs·core. We also subsample the fibers with a fixed sampling ratio 1/10, which reduces the computation to about 6000 hrs·core. Note that the tract distance calculation involves k-nearest neighbor search which is approximately linear time complexity with k-d tree. Since we implement the original GiTA on a large-scale computing array with thousands of CPUs, the computation time is reduced to a couple of days. By employing the GIFE framework, we reduce the computation time by over 90% to about 500 hrs·core, which is tractable for a small-size cluster with dozens of CPU cores.

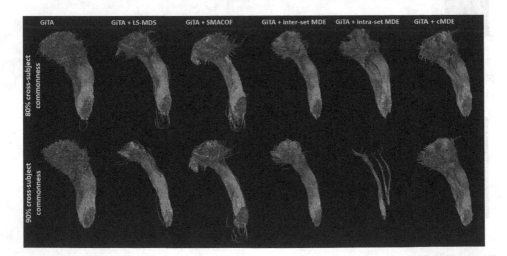

Fig. 3. Common optic radiation bundles. The common bundles from the original GiTA, GiTA + inter-set MDE and GiTA + cMDE generally do not contain spurious tracts.

We compare our cMDE method with the inter-set MDE and intra-set MDE as well as the original GiTA framework. The results are shown in Fig. 3. The results show that the original GiTA extracts the largest common bundles. However, the tracts appear to be a bit disorganized and this might be due to the inter-subject misalignment. We also observe that the GiTA + intra-set MDE framework failed to extract meaningful common bundle. Both of GiTA + inter-set MDE and GiTA + cMDE extract consistently highly organized bundles with increased organization by raising the commonness, while the bundles extracted by GiTA + cMDE are more anatomically complex and agreeable to the results of the original GiTA. This is an anticipated outcome of the better overlapped reference-target embedding of cMDE over the inter-set MDE. The computational time for the intra-set MDE for 1 subject is about 200 sec·core, and the inter-set MDE took about 0.05 sec·core, solving the optimal R^* and computing the final cMDE mapping took about 0.008 sec·core. We also compare with the MDU solved by LS-MDS and SMACOF in which we fix the reference embedding while iteratively updating only the target embedding for each target subject. However, LS-MDS and SMACOF generally require recalculating all the pairwise distances for the reference and target embeddings at each iteration, and each iteration of them took about 100 secs·core and both methods ran about 50 iterations before convergence. The results of LS-MDS are disorganized and sparse, and SMACOF gives a large amount of disorganized, hence invalid, common optic radiation fiber tracts.

4   Conclusion

In this work, we present a novel GIFE framework to achieve scalable GiTA. We also propose a novel method called MDE to achieve efficient and robust GIFE. The resultant method is highly scalable, parallelizable and robust to inter-subject misalignment. Real experiment shows clearly improved anatomical validity of the results of our proposed MDE method over other methods. This GIFE framework will be generally useful non-exclusively for common bundle reconstruction among all possible GiTA problems.

References

1. Maier-Hein, K.H., et al.: The challenge of mapping the human connectome based on diffusion tractography. Nat. Commun. 8(1), 1349 (2017)
2. Smith, R.E., Tournier, J.D., Calamante, F., Connelly, A.: SIFT: spherical-deconvolution informed filtering of tractograms. NeuroImage 67, 298–312 (2013)
3. Aydogan, D.B., Shi, Y.: Track filtering via iterative correction of TDI topology. In: Navab, N., Hornegger, J., Wells, W.M., Frangi, A.F. (eds.) MICCAI 2015. LNCS, vol. 9349, pp. 20–27. Springer, Cham (2015). https://doi.org/10.1007/978-3-319-24553-9_3
4. Wang, J., Aydogan, D.B., Varma, R., Toga, A.W., Shi, Y.: Topographic regularity for tract filtering in brain connectivity. In: Niethammer, M., Styner, M., Aylward, S., Zhu, H., Oguz, I., Yap, P.-T., Shen, D. (eds.) IPMI 2017. LNCS, vol. 10265, pp. 263–274. Springer, Cham (2017). https://doi.org/10.1007/978-3-319-59050-9_21

5. Wang, J., Shi, Y.: Gita: group-wise tractogram analysis. In: OHBM (2018)
6. Borg, I., Groenen, P.J.F.: Modern Multidimensional Scaling: Theory and Applications. SSS. Springer, New York (2005). https://doi.org/10.1007/0-387-28981-X
7. Weinberger, K.Q., Saul, L.K.: Unsupervised learning of image manifolds by semidefinite programming. Int. J. Comput. Vis. 70(1), 77–90 (2006)
8. O'Donnell, L.J., Westin, C.F.: Automatic tractography segmentation using a high-dimensional white matter atlas. IEEE Trans. Med. Imaging 26(11), 1562–1575 (2007)
9. Jianu, R., Demiralp, C., Laidlaw, D.: Exploring 3D DTI fiber tracts with linked 2D representations. IEEE Trans. Vis. Comput. Graph. 15(6), 1449–1456 (2009)
10. Drineas, P., Mahoney, M.W.: On the Nyström method for approximating a gram matrix for improved kernel-based learning. J. Mach. Learn. Res. 6, 2153–2175 (2005)
11. Lai, R., Osher, S.: A splitting method for orthogonality constrained problems. J. Sci. Comput. 58(2), 431–449 (2014)
12. Toga, A.W., Clark, K.A., Thompson, P.M., Shattuck, D.W., Van Horn, J.D.: Mapping the human connectome. Neurosurgery 71(1), 1–5 (2012)
13. Van Essen, D.C., et al.: The WU-Minn human connectome project: an overview. NeuroImage 80, 62–79 (2013)
14. Kammen, A., Law, M., Tjan, B.S., Toga, A.W., Shi, Y.: Automated retinofugal visual pathway reconstruction with multi-shell HARDI and FOD-based analysis. NeuroImage 125, 767–779 (2016)
15. De Leeuw, J.: Applications of convex analysis to multidimensional scaling. In: Recent Developments in Statistics (1977)

Multi-modal Brain Tensor Factorization: Preliminary Results with AD Patients

Göktekin Durusoy[1], Abdullah Karaaslanlı[1], Demet Yüksel Dal[1], Zerrin Yıldırım[2], and Burak Acar[1(✉)]

[1] Department of Electrical and Electronics Engineering, VAVlab, Boğaziçi University, Istanbul, Turkey
{goktekin.durusoy,abdullah.karaaslanli,demet.yuksel,acarbu}@boun.edu.tr
[2] Aziz Sancar Experimental Medical Research Institute, Department of Neuroscience, Istanbul University, Istanbul, Turkey
yildirimzerrin@gmail.com

Abstract. Global brain network parameters suffer from low classification performance and fail to provide an insight into the neurodegenerative diseases. Besides, the variability in connectivity definitions poses a challenge. We propose to represent multi-modal brain networks over a population with a single 4D brain tensor (**B**) and factorize **B** to get a lower dimensional representation per case and per modality. We used 7 known functional networks as the canonical network space to get a 7D representation. In a preliminary study over a group of 20 cases, we assessed this representation for classification. We used 6 different connectivity definitions (modalities). Linear discriminant analysis results in 90–95% accuracy in binary classification. The assessment of the canonical coordinates reveals Salience subnetwork to be the most powerful in classification consistently over all connectivity definitions. The method can be extended to include functional networks and further be used to search for discriminating subnetworks.

Keywords: Functional networks · Tensor factorization
Structural networks · Brain connectome · Alzheimer's Disease

1 Introduction

Brain has been known to be a network of cortical regions, yet until the relatively recent advances in magnetic resonance imaging (MRI), it was not possible to build network models of in-vivo brain. Current functional MRI (fMRI) and diffusion MRI (dMRI) technologies allow us to delineate functional (fNET) and structural network (sNET) models, collectively called the brain connectome, at a cortical parcellation scale. The analysis of these network models has the potential of shedding light on how the brain works, as well as the cause and progress of neurodegenerative diseases.

© Springer Nature Switzerland AG 2018
G. Wu et al. (Eds.): CNI 2018, LNCS 11083, pp. 29–37, 2018.
https://doi.org/10.1007/978-3-030-00755-3_4

The majority of the analysis approaches has been focused on the changes in the global network features, such as the clustering coefficient, the average path-length, the small-worldness index, etc. Despite their high sensitivity to abnormalities, these features are poor in classification and/or staging due to their global nature [1]. A more promising approach is to assess the changes in sub-networks of the connectome, towards which purely statistical techniques, such as Network Based Statistics (NBS) [2], have gained popularity. Another concern is the lack of standardized techniques for building connectomes, which introduces an unavoidable uncertainty on the derived conclusions.

We propose to use the powerful tensor factorization techniques to simultaneously address the aforementioned problems, namely sub-network based and multi-modal analysis of the connectome. We introduce the *Brain Tensor* (**B**-tensor) as a multi-dimensional multi-modal connectome representation and factorize it in terms of apriori known *canonical subnetworks*. In a preliminary study with Alzheimer's Disease patients, we demonstrate that the factorization coefficients not only have high discrimination power in a binary classification task, but also provide an insight into the most affected/discriminative canonical subnetworks. We conclude with a discussion on the potential extensions of **B**-tensor factorization.

2 Background

Initial efforts on network analysis have focused on global, and local characteristics in order to identify components of networks, and to assess similarities or differences between networks. Utilizing global features such as clustering coefficients, average path length, small-worldness is useful when a given network is compared with a reference network, or to examine differences of neural networks from different species. Briefly, global features look at the network as a whole and fail to identify local difference. On the other hand, local features such as local clustering coefficient, shortest path etc. have shown their significance when properties of individual components are examined [3].

NBS has been proposed to overcome the limitation of global assessment. It identifies statistically significantly different edges between two classes of networks. The identified edges are used to detect discriminative subnetworks. The distribution of sizes of these subnetworks is used to assign a p-value to the sub-network identified as discriminative between the positive classes. NBS is purely statistical, oversees the a priori known structure of brain.

Karahan et al. utilized coupled tensor factorization to fuse EEG, FMRI, and DTI data represented in 3rd, 2nd, and 3rd order tensors, respectively. Coupling is enforced over temporal and spatial domains for EEG/FMRI and over subjects for EEG/DTI. Two different representation is used for EEG where first one is time-varying and the other one is subject-varying EEG [4]. However, in this research, a priori is not used and multimodal structural and functional representation of brain networks have not been considered. Utilized tensors do not represent network sets row signals/spectra with a spatial and/or subject dimension. Williams et al. generate a 3rd order tensor model by using a trial-structured neural data with dimensions represent neurons, time and trials. With the help of TCA (Tensor Component Analysis), they have managed to decompose this tensor into three interpretable factors (neuron factors, temporal factors, trial factors). In addition, TCA has been utilized for dimension reduction [5].

We propose a 4th order **B**-tensor that represents multi-modal networks (structural and/or functional) over a population. A modality is defined as a network construction method independent from data source (i.e. structural or functional data). Decomposition of **B**-tensor over a priori known subnetwork is studied.

3 Method

3.1 B-Tensor Construction

For a multi-modal (R-modal) connectome defined over $I \times J$ nodes (cortical parcels) for a population of K cases, the 4th order B-tensor ($\mathbf{B} \in \mathbb{R}^{K \times I \times J \times R}$) is defined. In this work, we used 6 variants of sNET definitions, hence $R = 6$, over a population of 20 cases, hence $K = 20$. We used the 148-parcel Destrieux atlas [6], hence $I = J = 148$.

Following the co-registration of T1-weighted MRI and dMRI volumes, the T1-weighted MRI volume is parcellated using FreeSurfer[1], into 148 parcels which are used to define the 148 nodes ($\{V_i\}$) of the sNETs. Diffusion tensor (DT) volume is computed from diffusion weighted MRI (DWI) using an in-house software built upon the MITK platform[2]. Fiber tracts are constructed using the 4th-order Runge-Kutta (RK4) deterministic tractography algorithm [7] with minimum fractional anisotropy (FA) set to 0.15, stepsize set to 0.7 mm ($\approx$ half the voxel size), minimum fiber length set to 14 mm and the maximum curvature set to 35°. RK4 was initiated from 30 randomly selected seeds per voxels with $FA > 0.15$.

In order to construct the sNETs, each fiber ($\{f_k\}$) is associated with the nodes in the vicinity of its end-points using a symmetric 3D Gaussian kernel with a standard deviation (σ) of 0.155 mm, centered at the fiber endpoints. σ is optimized by minimizing the integrated square error (ISE), as described in [8]. The numeric volumetric integral of the Gaussian kernel positioned at one of the end points of f_k, within the node V_i (and up to a radial distance of 2σ) is used as the fiber-parcel/node association and is denoted by W_{ik}.

Six different sNETs are constructed using 6 different structural connectivity (network edge weight) definitions, C_{ij}, between pairs of nodes, (V_i, V_j), as follows:

$$C_{ij} = \sum_k W_{ik}W_{jk} : \text{Weighted Connectivity} \tag{1}$$

---

[1] https://surfer.nmr.mgh.harvard.edu/.
[2] http://mitk.org/wiki/MITK.

$$C_{ij}^N = \frac{2 \times \frac{C_{ij}}{\mathcal{V}}}{V_i + V_j} : \text{Normalized Connectivity} \tag{2}$$

$$C_{ij}^{stat} = \frac{1}{C_{ij}} \sum_k W_{ik} W_{jk} \times \Psi(FA(f_k(t))) \ , \ \forall C_{ij} \neq 0$$

$$\text{FA-based Connectivities} \tag{3}$$

where $\mathcal{V}$ is the voxel volume in mm^3, V_i and V_j are the volumes of corresponding parcels, and Ψ represents the statistics operator ($\in\{$min, max, mean, median$\}$) operating over the fiber parametrized by t.

3.2   B-Tensor Factorization

The CP factorization of $\mathbf{B}$ is given as [9,10],

$$\mathbf{B}_{k,i,j,r} \approx \sum_{q=1}^Q \mathbf{A}_{k,q} \mathbf{C}_{i,q} \mathbf{D}_{j,q} \mathbf{E}_{r,q} \tag{4}$$

where the decomposition is performed over Q (free parameter) factors. Each one of the components represents the factorization of a single dimension of $\mathbf{B}$ over the Q factors. While $\mathbf{A}$ and $\mathbf{E}$ are associated with the individual cases and the sNET definitions, $\mathbf{C}$ and $\mathbf{D}$ are solely associated with the network topology. Hence, we combined $\mathbf{C}$ and $\mathbf{D}$ into a single component that represents network topologies across Q factors, while $\mathbf{A}$ and $\mathbf{E}$ are merged to represent per case per connectivity (per-modality) represented in terms of those network topologies. Namely,

$$\mathbf{B}_{k,i,j,r} \approx \sum_{q=1}^Q \mathbf{M}_{k,r,q} \mathbf{G}_{i,j,q} = \sum_{q=1}^Q \left(\sum_s \mathbf{A}_{k,s} \mathbf{E}_{r,s} \mathbf{I}_{q,s} \right) \left(\sum_s \mathbf{C}_{i,s} \mathbf{D}_{j,s} \mathbf{I}_{q,s} \right) \tag{5}$$

where $\mathbf{I}_{q,s} = \delta_{qs}$, i.e. the identity matrix and $s \in \{1, 2, \cdots, Q\}$. This allows us to decouple the network topology from the cases and the modalities (connectivity definitions). Thus, we can work with case and modality independent network topologies, namely the *canonical subnetworks*, $\mathbf{G}$. With a further simplification, we constrained $\mathbf{G}$ to be a binary valued tensor representing the apriori known (fixed) canonical subnetworks. Following Yeo et al., we defined 7 canonical subnetworks, namely the visual, the somatomotor and auditory, the dorsal attention, the salience, the limbic, the frontoparietal and the default mode subnetworks [11]. They are expressed in terms of the node definitions of $\mathbf{B}$ and numbered from 1 to 7, respectively. $\mathbf{G}$ is assumed known and fixed for the rest. $\mathbf{M}$, on the other hand, represents the factorization coefficients over the canonical dimensions (subnetworks). This gives us 7D representations per case and per modality as $\mathbf{F}_{k,r} \in \mathbb{R}^7$.

Following [12], Eq. 5 can be matricised as

$$\mathbf{B}_{(1,4;2,3)} = \mathbf{M}_{(3;1,2)}^T \mathbf{G}_{(3;1,2)} \tag{6}$$

where $\mathbf{M}_{(3;1,2)} = \mathbf{M}_{(3)} = (\mathbf{E} \odot \mathbf{A})^T$ and $\mathbf{G}_{(3;1,2)} = (\mathbf{G}_{(3)} = \mathbf{D} \odot \mathbf{C})^T$.[3] Fixing $\mathbf{G}$ as described above, we can solve for $\mathbf{M}$ using any matrix inversion technique, such as QR decomposition, to get $\mathbf{M}_{(3)} = \mathbf{G}_{(3)}^{-T}\mathbf{B}_{(1,4;2,3)}^T$. $\mathbf{G}_{(3)}^{-T}$ is computed once and used throughout the analysis. Finally, the computed $\mathbf{M}_{(3)}$ is tensorized back to its original form to get the estimated $\mathbf{M}$ as a 3rd order tensor representing per case, per connectivity unconstrained real-valued factorization coefficients.

4 Experiments

4.1 Data

T1-weighted MRI and dMRI images were acquired by using a Philips Achieva 3.0T X scanner with a 32-channel head coil from 7 AD patients and 13 controls with written consent. The AD patients were diagnosed by means of standard clinical evaluation tests by a team of expert neurologists. We used 3D FFE (Fast Field Echo) pulse sequence with multi-shot TFE (Turbo Field Echo) imaging mode for T1-weighted MRI. The acquisition parameters were TE/TR = 3.8ms/8.3ms, flip-angle = 8°, SENSE reduction 2 (Foot-Head), FOV = $220(RL) \times 240(AP)\,\text{mm}^2$, voxel size = $1.0 \times 1.0 \times 1.0\,\text{mm}^3$ and number of slices = 180. dMRI were acquired with a maximum gradient strength of 40 mT/m, and 200 mT/m/ms slew rate, using a single-shot, pulse-gradient spin echo (PGSE), echo planar imaging (EPI) sequence. The acquisition parameters were FOV= $200 \times 236\,\text{mm}^2$, 2.27 mm isotropic voxel size, 112×112 reconstruction matrix, 71 slices and $TE/TR = 92\,\text{ms}/9032\,\text{ms}$. 120 diffusion weighting gradient directions were used at various b-values between $3000 - 0\,\text{s/mm}^2$.

4.2 Analysis and Results

The separation of the AD patients and the controls in the 7D canonical space of subnetworks was assessed by Linear Discriminant Analysis(LDA) [13]. Binary classifiers are trained for each one of the 6 connectivity definitions in the corresponding 7D space ($\mathbf{F_r}$) and the training accuracies are measured. We also computed the unit normal of separating hyperplanes, i.e. the *canonical axis*, along which the separation of the two groups is maximized. Acc_B represents the binary classification accuracy. We also trained a separate binary classifier using LDA on the standard global network parameters (global clustering coefficients [14], average shortest path length [15], small-worldness index [16]). Acc_{glb} represents the classification accuracy using these global parameters. The results are given in Table 1 together with the corresponding canonical axes for the B-tensor analysis. All accuracies were above 90%, with the weighted and normalized connectivity definitions being the best performing ones for **B**-tensor. The corresponding classification accuracies using global connectome features in LDA are between 75%–90%.

---

[3] $\odot$ denotes the Khatri-Rao product.

Table 1. Classification accuracies of LDA classifier for different connectivity definitions (modalities) and the associated canonical axes.

Conn	Canonical Axis	Acc_B	Acc_{glb}
C_{ij}	[0.012 −0.010 −0.377 −0.887 0.211 −0.159 −0.022]	0.95	0.8
C_{ij}^N	[−0.326 −0.130 −0.530 −0.636 0.426 0.054 −0.073]	0.95	0.85
C_{ij}^{min}	[−0.224 −0.064 −0.351 −0.509 0.401 0.448 −0.446]	0.90	0.75
C_{ij}^{max}	[−0.352 0.231 −0.345 −0.703 0.299 0.056 0.340]	0.90	0.90
C_{ij}^{mean}	[−0.352 −0.287 −0.298 −0.714 0.347 0.046 0.215]	0.90	0.80
C_{ij}^{median}	[−0.333 −0.296 −0.286 −0.741 0.347 0.0461 0.215]	0.90	0.75

The components of the canonical axes unit vectors provide an insight with regard to the relevance of the corresponding canonical subnetwork in discriminating the AD patients from the controls. The higher the absolute value of a component of a canonical axes, the more important that canonical dimension (subnetwork) is in discriminating the two groups. Figure 1 shows the mean and standard deviation of the magnitude of each component computed over all modalities.

In order to compare the connectivity definitions with regard to their discriminating power, we ordered the canonical subnetworks based on the mean values given in Fig. 1 and computed the classification accuracy of LDA using the top K ($\in [1, 7]$) canonical subnetworks, separately for each connectivity definition. Figure 2 shows the results. In general, increasing the dimension of the canonical space improves the accuracy, except for the 2D case (i.e. using the dorsal

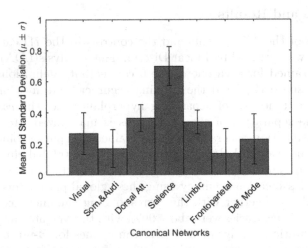

Fig. 1. Mean and standard deviations of the canonical axes' components computed over all modalities. The 4th subnetwork (Salience subnetwork) is consistently observed to be the most important one.

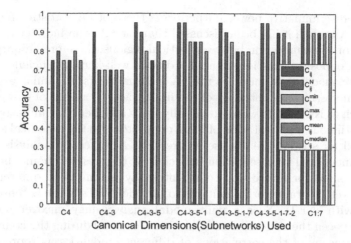

Fig. 2. Accuracy of LDA for all modalities using top K ($\in [1, 7]$) canonical dimensions (subnetworks). The weighted connectivity definition performed the best almost unanimously where as the FA-statistics based connectivities performed relatively poorly in general.

attention and the salience networks only). However, the weighted connectivity definition performed the best almost unanimously where as the FA-statistics based connectivities performed relatively poorly in general.

5   Discussion

The B-tensor factorization allows us to represent the multi-modal (multi-connectivity) brain connectome in a canonical space of subnetworks with an intuitive interpretation. The AD patients are clearly separated from the controls in this space. The salience network is consistently observed to be the most important subnetwork among the 7 subnetworks used. The dorsal attention and the limbic subnetworks seem to be the second most important networks whereas the fronto-parietal network is the least important one. Although this result seems to be counter intuitive as the memory loss (primary function of limbic subnetwork) is the major symptom of AD, there is also evidence supporting our findings [17]. Furthermore, the AD cases in our dataset are described as early stage AD by our collaborating neurologists, which may also explain the observed importance of the salience subnetwork. These preliminary results are limited by the fixed definition of the canonical subnetworks. However, the B-tensor factorization framework can be utilized for searching for discriminative subnetworks. This would provide a further insight into the causes and progression mechanism of neurodegenerative diseases.

The assessment of different connectivity definitions within the aforementioned canonical space reveals that the weighted and normalized connectivity definitions' discriminative power outperforms those of FA-statistics based connectivities. Although it has been discussed that the FA is an indirect measure of the quality of communication between cortical regions, these preliminary results suggest the opposite. This is potentially due to well-known deficiencies of diffusion tensor model that underlies the FA measurements. A similar assessment using the microstructural integrity of the fiber tracts by means of compartment models, such as NODDI [18], can potentially show a higher discriminative power.

The continuous representation of brain connectome in the canonical space can also be used for staging the disease progression. A regression analysis between this low-dimensional representation and clinical test results should be carried out, which is left for future work. Current study is limited to different sNET definitions, yet the framework is suitable to include fNETs in the B-tensor simultaneously with sNETs. Such a multi-modal analysis may uncover non-trivial relations between the structural and functional changes during the course of the disease, by means of the correlations of different modalities as represented in the canonical space. A fundamental limitation of the present study is the small dataset size which might have caused an overfitting of the LDA classifier. Further experiments on a larger dataset are due to arrive at stronger conclusions.

6   Conclusion

We have presented a novel tensor based multi-modal representation of brain connectome and described how it can be factorized to get a continuous, low-dimensional representation in a canonical space offering an intuitive understanding of neurodegenerative diseases' causes and progression. Preliminary results with a small cohort of AD patients and controls revealed high classification accuracy and identified the salience subnetwork as the most discriminative network component among the 7 known.

Future work will include the experiments with a larger cohort, the extension of the model to joint analysis of structural and functional networks, the assessment of the canonical representation as a disease staging/monitoring biomarker and developing a canonical subnetwork search strategy optimized for classification/regression accuracy.

Acknowledgement. This work was in part supported by the Turkish Ministry of Development under the TAM Project number DPT2007K120610, and in part by TUBITAK-ARDEB project number 114E053.

References

1. Fornito, A., Zalesky, A., Bullmore, E.T.: Fundamentals of Brain Network Analysis. Academic Press, San Diego (2016)
2. Zalesky, A., Fornito, A., Bullmore, E.T.: Network based statistic: identifying differences in brain networks. Neuroimage **53**(4), 1197–1207 (2010)
3. Kaiser, M.: A tutorial in connectome analysis: topological and spatial features of brain networks. Neuroimage **57**, 892–907 (2011)

4. Karahan, E., Rojas-Lopez, P.A., Bringas-Vega, M.L., Valdes-Hernandez, P.A., Valdes-Sosa, P.A.: Tensor analysis and fusion of multimodal brain images. Proc. IEEE **103**, 1531–1559 (2015)
5. Williams, A.H., et al.: Unsupervised discovery of demixed, low-dimensional neural dynamics across multiple timescales through tensor component analysis. Neuron (2018)
6. Destrieux, C., Fischl, B., Dale, A., Halgren, E.: Automatic parcellation of human cortical gyri and sulci using standard anatomical nomenclature. Neuroimage **53**(1), 1–15 (2010)
7. Tench, C.R., Morgan, P.S., Wilson, M., Blumhardt, L.D.: White matter mapping using diffusion tensor MRI. Magn. Reson. Med. **47**(5), 967–972 (2002)
8. Moyer, D., Gutman, B.A., Faskowitz, J., Jahanshad, N., Thompson, P.M.: Continuous representations of brain connectivity using spatial point processes. Med. Image Anal. **41**, 32–39 (2017)
9. Kolda, T.G., Bader, B.W.: Tensor decompositions and applications. SIAM Rev. **51**(3), 455–500 (2009)
10. Kiers, H.A.L.: Towards a standardized notation and terminology in multiway analysis. J. Chemometrics **14**, 105–122 (2000)
11. Yeo, B.T., et al.: The organization of the human cerebral cortex estimated by intrinsic functional connectivity. J. Neurophysiol. **106**(3), 1125–65 (2011)
12. Kolda, Tamara G.: Multilinear operators for higher-order decompositions, Technical report 2081. SANDIA, Albuquerque, NM (2006)
13. Fisher, R.A.: The use of multiple measurements in taxonomic problems. Ann. Eugenics **7**, 179–188 (1936)
14. Onnela, J.P., Saramäki, J., Kertész, J., Kaski, K.: Intensity and Coherence of motifs weighted complex networks. Phys. Rev. E. **71**(6), 065103 (2005)
15. Boccaletti, S., Latora, V., Moreno, Y., Chavez, M., Hwang, D.-U.: Complex networks: structure and dynamics. Phys. Rep. **424**(4–5), 175–308 (2006)
16. Humphries, M.D., Gurney, K.: Network small-world-ness: a quantitative method for determining canonical network equivalence. PLOS ONE **3**(4), 110,04 (2008)
17. Seeley, W.W., Crawford, R.K., Zhou, J., Miller, B.L., Greicius, M.D.: Neurodegenerative diseases target large-scale human brain networks. Neuron **62**, 42–52 (2009)
18. Zhang, H., Schneider, T., Wheeler-Kingshott, C.A., Alexander, D.C.: NODDI: practical in vivo neurite orientation dispersion and density imaging of the human brain. Neuroimage **61**, 1000–1016 (2012)

Intact Connectional Morphometricity Learning Using Multi-view Morphological Brain Networks with Application to Autism Spectrum Disorder

Alaa Bessadok[1,2] and Islem Rekik[1(✉)]

[1] BASIRA Lab, CVIP Group, School of Science and Engineering, Computing,
University of Dundee, Dundee, UK
irekik@dundee.ac.uk
[2] National Engineering School of Gabes, Gabes, Tunisia
http://www.basira-lab.com/

Abstract. The morphology of anatomical brain regions can be affected by neurological disorders, including dementia and schizophrenia, to various degrees. Hence, identifying the morphological signature of a specific brain disorder can improve diagnosis and better explain how neuroanatomical changes associate with function and cognition. To capture this signature, a landmark study introduced, brain *morphometricity*, a global metric defined as the proportion of phenotypic variation that can be explained by brain morphology derived from structural brain MRI scans. However, this metric is limited to investigating morphological changes using low-order measurements (e.g., regional volumes) and overlooks how these changes can be related to each other (i.e., how morphological changes in region A are influenced by changes in region B). Furthermore, it is derived from a *pre-defined* anatomical similarity matrix using a Gaussian function, which might not be robust to outliers and constrains the locality of data to a fixed bandwidth. To address these limitations, we propose the *intact connectional brain morphometricity* (ICBM), a metric that captures the variation of *connectional* changes in brain morphology. In particular, we use *multi-view morphological brain networks* estimated from multiple cortical attributes (e.g., cortical thickness) to learn an intact space that first integrates the morphological network views into a unified space. Next, we *learn* a multi-view morphological similarity matrix in the intact space by adaptively assigning neighbors for each data sample based on local connectivity. The learned similarity capturing the shared traits across morphological brain network views is then used to derive our ICBM via a linear mixed effect model. Our framework shows the potential of the proposed ICBM in capturing the connectional neuroanatomical signature of brain disorders such as Autism Spectrum Disorder.

© Springer Nature Switzerland AG 2018
G. Wu et al. (Eds.): CNI 2018, LNCS 11083, pp. 38–46, 2018.
https://doi.org/10.1007/978-3-030-00755-3_5

1   Introduction

Brain disorders affect the brain construct on multiple levels including neural activity quantified using functional magnetic resonance imaging (MRI) and brain tissue morphology measured using structural T1-weighted MRI. While several studies focused on identifying the functional signature (or fingerprint) of brain disorders [1–3], a few works investigated the morphological fingerprint of a specific brain disorder (Alzheimer's disease, Autism Spectrum Disorder, Parkinson's disease). To fill this gap, [4] proposed a statistical metric called brain 'morphometricity' (BM) that describes the associations between brain morphology and multiple risk factors such as age and gender. Using structural MRI, volumetric measurements of noncortical structures and thickness measurements of cortical regions were generated. To capture the similarity between brain morphologies of brains drawn from distinct groups (e.g., normal controls and ASD patients), they computed a similarity matrix for each of these measurements separately, and then averaged them to produce the global anatomical similarity matrix. Ultimately, a Linear Mixed Effect model (LME) was applied to estimate the variance captured by the similarity matrix to unravel the morphological signature of a specific phenotypic trait (e.g., clinical diagnosis).

However, the proposed morphometricity metric is limited to investigating morphological changes using low-order measurements (e.g., regional volumes) and overlooks how these changes can be related to each other (i.e., how morphological changes in region A are influenced by changes in region B). In other words, it does not look at morphological connectivity of anatomical regions of interest (ROIs), where a morphological connection quantifies the (dis)similarity in shape between two brain ROIs –i.e., how their morphologies are related. This can be modeled using multi-view morphological brain networks (MBN) as proposed in [5–8]. These showed great potential for brain disorder diagnosis [5–7] and morphological connectional biomarker identification [8] using supervised [5,6] or unsupervised learning [7] techniques trained on structural T1-weighted MRI data. More importantly, each view-specific MBN models the relationship *in morphology* between brain regions using a specific measurement (e.g., curvature).

To fill this gap, we unprecedentedly propose to use multi-view MBNs for *'connectional brain morphometricity'* (CBM) estimation. We note that in the landmark work [4] of BM, the similarity matrix is computed using a pre-defined similarity function such as Gaussian metric, which (i) may not be robust to outliers, (ii) may not handle well multi-view data drawn from multiple sources, and (iii) may fail to capture data sample distributions with varying bandwidths. To address these limitations, we propose to *learn* the data similarity matrix by levering the intact-awareness similarity learning model developed in [9]. More precisely, the proposed approach aims to recover an intact space [10] that captures the complementarity between multiple data views. A practical example of this is the medical diagnosis of neurological diseases, such as dementia. Each morphological feature (e.g., cortical thickness) alone captures insufficient information and thus cannot comprehensively describe the brain atrophy, which can only be fully recovered by integrating all the features. To leverage the complementary of multi-view MBNs, we propose a novel *intact connectional brain morphometricity* (ICBM) learning framework to identify the connectional morphology-driven fingerprint of specific traits. Specifically, we use the intactness-aware similarity learning method [9] to estimate the similarity that has the maximum dependence with its intact space, where shared traits across views are well captured. First, we learn the complementarity between different MBNs by constructing an intact connectomic space. Within a joint framework, we simultaneously *learn* a multi-view morphological similarity matrix in the intact space by adaptively assigning neighbors for each data sample based on local connectivity. The learned similarity capturing the shared traits across morphological brain network views is then used to derive our ICBM via a linear mixed effect model. The main contributions of our work can be summarized as follows:

- We propose to learn a morphological intact space that models the complementarity between different morphological brain networks by integrating them in one space, thereby catching partial information from each individual view.
- We learn the multi-view morphological similarity matrix that is in harmony with the morphological intact space of multi-view MBNs.
- We introduce the intact connectional brain morphometricity, a metric that could reveal novel insights into morphological connectivity fingerprinting brain disorders.

2 Intact Connectional Brain Morphometricity Learning

In the following, we present the main steps of our intact connectional brain morphometricity (ICBM) learning framework. To clarify the reading, we summarized the major mathematical notation in Table 1. Figure 1 illustrates the proposed pipeline for estimating the intact connectional morphometricity in three major steps: (1) construction of multi-view morphological networks, (2) learning of the intact multi-view similarity matrix, and (3) estimation of the ICBM using LME model.

Multi-view Morphological Network. Inspired by the foundational works of [7,8], we define a morphological brain network $\mathbf{V}$ as a graph comprising a set of nodes, each node representing a brain ROI. The connection between two nodes quantifies the dissimilarity in shape between two ROIs i and j by computing the absolute difference between ROI-based average morphological measurements (e.g., mean curvature). By diversifying the morphological measurements, we generate a set of MBNs $\mathcal{M}_{\mathbf{v}} = \{V^1, V^2, \ldots, V^k\}$, each capturing a specific view of the morphological brain construct. Since each MBN can be defined as a symmetric matrix, we only vectorize the off-diagonal upper triangular part to generate a feature vector $\mathbf{F}_s^k$ for each subject s and each view k (Fig. 1-A).

Intact Morphological Similarity Learning. This step is the core of our framework as it describes the connectional similarity between the morphological views. Basically, we first propose to learn an intact space that represents the

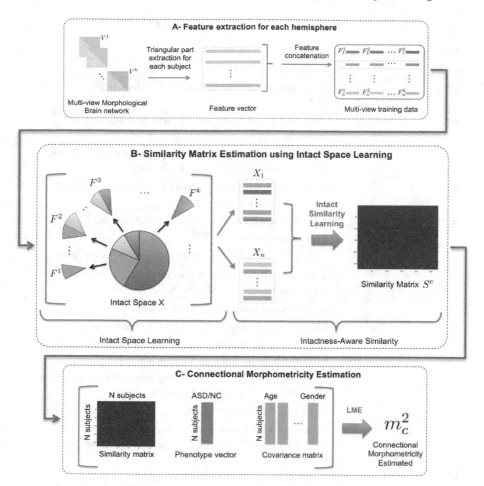

Fig. 1. *Proposed framework for intact connectional brain morphometricity (ICBM) learning.* (A) Feature extraction from different multi-view morphological brain networks, each driven from a specific morphological brain measurement (e.g., curvature). Multi-view feature vectors are concatenated to create a multi-view training matrix including all subjects. (B) Intact similarity matrix construction. We learn an intact connectomic space, which captures the complementarity between all views $\{\mathbf{F}^k\}$, and where the intact similarity matrix $\mathbf{S}$ is jointly learned. (C) ICBM estimation. Given the learned similarity matrix along with the phenotype vector (e.g., subject label as normal control or autistic) and the population covariance matrix, we compute ICBM using linear mixed effect (LME) model.

complementary information of multiple views. As reported in [10], an individual view is insufficient for learning, thus integrating multiple views is necessary to learn a comprehensive representation of the data. Given specific views $\mathbf{V^i}$ and $\mathbf{V^j}$ generated from the intact space $\mathbf{X}$, the view insufficiency can be expressed by $\mathbf{I}(\mathbf{X}; \mathbf{V}^j | \mathbf{V}^i)$ that measures how much information is shared between the intact

Table 1. Major mathematical notations used in this paper

Mathematical notation	Definition
n	number of subjects
$\mathbf{V}^k$	brain network of the k view of subject n in $\mathbb{R}^{n_r \times n_r}$
$\mathcal{M}_\mathbf{v}$	set of subject-specific multi-view morphological networks
$\mathbf{F}^k$	matrix including features vectors extracted from the kth brain network view k of all subjects
$\mathbf{F}^k_s$	feature vector extracted from the brain network of the k-th view for subject s
$\mathbf{X}_n$	a sample in the intact space X represented by K feature vectors F^k in $\mathbb{R}^{d^k_f}$
	where d^k_f is the feature dimension of the k-th view
$\mathbf{W}^k$	a mapping function of a specific feature view $\mathbf{F}^k$, representing all subjects in $\mathbf{X}$
$\mathbf{S}^c$	connectional similarity matrix in $\mathbb{R}^{n \times n}$
$\mathbf{m}^2_c$	learned intact connectional brain morphometricity
y	the phenotype vector that describes the clinical state of samples (e.g., healthy or disordered)
$\mathbf{\Sigma}$	the covariance matrix that contains data of covariate variables such as age and gender
$\mathbf{f}_e$	the LME fixed effect vector
$\mathbf{r}_e \sim N(0, v_a \mathbf{S}^c)$	a random effect vector resulting from a zero-mean multivariate Gaussian distribution with a covariance matrix $\mathbf{S}^c$
ϵ	the noise vector with variance v_e

space $\mathbf{X}$ and the newly generated view $\mathbf{V}^\mathbf{j}$ given the known view $\mathbf{V}^\mathbf{i}$. Given, multiple views $\mathcal{M}_\mathbf{v}$ generated from the complete intact space $\mathbf{X}$, the information obtained to learn $\mathbf{X}$ is measured by:

$$\mathbf{I}(\mathbf{X}; \mathbf{V}^1, \mathbf{V}^2, \ldots, \mathbf{V}^k) = \sum_{i=1}^{k} \mathbf{I}(\mathbf{X}; \mathbf{V}^{i-1}, \mathbf{V}^{i-2}, \ldots, \mathbf{V}^1) \tag{1}$$

Thus, learning $\mathbf{X}$ can be formulated as a minimization problem based on Eq. 1 so that $\mathbf{X} = \min_\mathbf{X} L(X; V^1, V^k)$ where $L(.)$ is the loss function $l(.)$ over the samples on different views. Considering $\mathbf{W}^k$ a mapping function of a specific feature view $\mathbf{F}^k$ representing all subjects in the intact space $\mathbf{X}$, the intact space $\mathbf{X}$ learning is formulated as follows:

$$\min_{\mathbf{X}, \mathbf{W}^k} \frac{1}{\mathbf{K}} \sum_{k=1}^{K} \|\ \|^2_F + \lambda_1 \|\ \|^2_F \tag{2}$$

where $\lambda_1 \|\|_F^2$ is a regularization term used to penalize the intact space $\mathbf{X}$ and λ_1 is a non-negative parameter.

Following the learning of the intact space $\mathbf{X}$, one can learn an intact similarity matrix $\mathbf{S}^c$ between subjects across views by maximizing its dependence with the intact space $\mathbf{X}$. This results in connecting the data points based on their locality –i.e., only the nearest neighbors observations of a specific point can be connected to this point rather than all other observations. Hence, the *multi-view morphological similarity learning* can be formulated as follow:

$$\min_{\mathbf{S}^c} \lambda_2 \sum_{i=1}^{n} \sum_{j=1}^{n} \|_1 \mathbf{S}_{ij}^c + \gamma \|_F^2 \tag{3}$$

where $\gamma \|\|_F^2$ is used to prevent $\mathbf{S}^c$ from converging to identity matrix. λ_2 and γ are non-negative parameters. Additionally, in order to handle noisy samples, we adopted the l_1 distance instead of l_2.

Since the connectional similarity matrix $\mathbf{S}^c$ is derived from the intact connectomic space $\mathbf{X}$, we combine both models of Eqs. 2 and 3 into a joint alternating optimization framework where the learning of the intact space is influenced by the learning of the similarity matrix and vice versa:

$$\min_{\mathbf{X}, \mathbf{W}^k, \mathbf{S}^c} \frac{1}{K} \sum_{k=1}^{K} \|_F^2 + \lambda_2 \sum_{i=1}^{n} \sum_{j=1}^{n} \|\mathbf{X}_i - \mathbf{X}_j\|_1 \mathbf{S}_{ij}^c + \gamma \|_F^2 \tag{4}$$

Intact Connectional Brain Morphometricity Estimation. Next, we propose to use the learned intact morphological similarity matrix $\mathbf{S}^c$ to estimate the intact connectional morphometricity. Specifically, we are using the Restricted Maximum Likelihood (ReML) [11] to fit the Linear Mixed Effect (LME) model described as follows:

$$\mathbf{y} = \mathbf{\Sigma} * \mathbf{f}_e + \mathbf{r}_e + \epsilon, \tag{5}$$

where $\mathbf{y}$ denotes the phenotype vector that describes the clinical state of samples (e.g., healthy or disordered subject), $\mathbf{\Sigma}$ is the covariance matrix that contains data of covariate variables such as age and gender, $\mathbf{f}_e$ is the fixed effect vector, $\mathbf{r}_e \sim N(0, v_a \mathbf{S}^c)$ is a random effect vector resulted from a zero-mean multivariate Gaussian distribution with a covariance matrix $\mathbf{S}^c$, and ϵ denotes the noise vector with variance v_e. We then define the intact connectional brain morphometricity m_c as:

$$m_c = \frac{v_a}{v_a + v_e} = \frac{v_a}{v_c} \tag{6}$$

where v_a is the variance captured by $\mathbf{S}^c$ and v_c is the phenotypic variance. The proposed ICBM can thus described as the proportion of phenotypic variation that can be explained by morphological brain connectivity.

3   Results and Discussion

Data Parameters. We evaluate the proposed framework on 341 subjects (155 ASD and 186 NC) from Autism Brain Imaging Data Exchange (ABIDE)[1], each represented using four morphological brain networks constructed using the following cortical measurements in the right and left hemispheres: mean maximum principal curvature, mean cortical thickness, mean sulcal depth, mean of average curvature. For more details about MBN construction strategy, we kindly refer the reader to [6,8]. Three parameters were tuned using grid search: the dimension of the intact space, λ_2 is a non-negative trade-off parameter and n_k is the number of nearest neighbor of a specific sample in $\mathbf{X}$. Specifically, using a grid search strategy we tuned one parameter by fixing the others using 5-fold cross-validation for the left and the right hemispheres, independently.

Estimating ICBM Using Different Combinations of Brain Network Views. Given our 4 brain network views, we first constructed all possible combinations using 2, 3, and 4 views, respectively. This allows to investigate the ICBM using different combinations of views as mapped onto the intact space. For instance, using two views, we generate C_4^2 ICBMs, each for a specific pair of views. Next, we report in Fig. 2 the average ICBM across all pairings along with the standard deviation. For comparing the estimated intact connectional brain morphometricity across hemispheres, we report in Fig. 2-A ICBM estimates when tuning the parameters for the left hemisphere (LH) and then fixing them for the right hemisphere (RH), whereas in Fig. 2-B, the ICBM parameters are tuned using the RH.

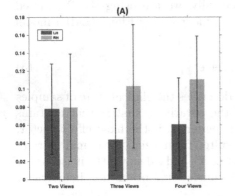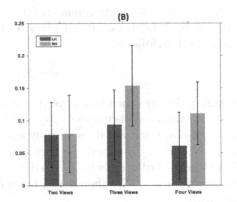

Fig. 2. Intact connectional brain morphometricity (ICBM) estimates using three different combinations of brain four views: mean maximum principal curvature, mean cortical thickness, mean sulcal depth, mean of average curvature. (A) ICBM estimated while fixing the model parameters using the left hemisphere (LH). (B) ICBM estimated while fixing the model parameters using the right hemisphere (RH). Blue bars display ICBM for the LH and orange bars display ICBM for the RH.

---

[1] http://fcon_1000.projects.nitrc.org/indi/abide/.

Figure 2 shows the association between multi-view morphological networks and ASD, assessed using the ICBM. Specifically, our preliminary analyses indicate that this particular clinical condition is not significantly morphometric since all intact connectional brain morphometricity estimates were smaller than 0.8 as explained in [4]. Figure 2 also shows that the right hemisphere (orange bars) is more morphometric than the left hemisphere on a 'connectional' level. This is in line with the findings of [7], where MBNs derived from the right hemisphere produced the best classification accuracy in distinguishing between ASD and NC subjects, which might indicate that right hemispheric connectional features have more discriminative power than the left hemisphere when leveraging high-order morphological information such as correlation between cortical measurements. We also note that both Fig. 2-A and B exhibit similar trends where the estimated of ICBM for RH is higher than LH for three- and four-view based combinations. As for two-view based combination, we note that ICBN is somewhat invariant across cortical hemispheres.

4   Conclusion

In this work, we introduced the intact connectional brain morphometricity, a metric that is learned using multi-view morphological brain network data for identified the connectional morphometric fingerprint of a specific trait (e.g., autism spectrum disorder). Our preliminary results revealed that autism is not significantly morphometric on a *connectional* level. However, we found that the right hemisphere is more morphometric than the left one. In our future work, we will evaluate the proposed ICBM learning framework on other disordered datasets (e.g., dementia). It would be also interesting to compare conventional brain morphometricity [4] to the connectional one.

References

1. Collin, G.: The connectomic blueprint of Schizophrenia. Ph.D thesis (2015)
2. Finn, E.S., Shen, X., Scheinost, D., Rosenberg, M.D., Huang, J., Chun, M.M., Papademetris, X., Constable, R.T.: Functional connectome fingerprinting: identifying individuals using patterns of brain connectivity. Nat. Neurosci. **18**, 1664 (2015)
3. Imperiale, F., Agosta, F., Canu, E., Markovic, V., Inuggi, A., Jecmenica-Lukic, M., Tomic, A., Copetti, M., Basaia, S., Kostic, V.: Brain structural and functional signatures of impulsive-compulsive behaviours in Parkinson's disease. Mol. Psychiatry **23**, 459 (2018)
4. Sabuncu, M.R., et al.: Morphometricity as a measure of the neuroanatomical signature of a trait. Proc. Nat. Acad. Sci. **113**, E5749–E5756 (2016)
5. Lisowska, A., Rekik, I., Initiative, A.D.N., et al.: Pairing-based ensemble classifier learning using convolutional brain multiplexes and multi-view brain networks for early dementia diagnosis. In: International Workshop on Connectomics in Neuroimaging, pp. 42–50 (2017)

6. Lisowska, A., Rekik, I.: Joint pairing and structured mapping of convolutional brain morphological multiplexes for early dementia diagnosis. Brain connectivity (2018)
7. Soussia, M., Rekik, I.: High-order connectomic manifold learning for autistic brain state identification. In: International Workshop on Connectomics in Neuroimaging, pp. 51–59 (2017)
8. Mahjoub, I., Mahjoub, M.A., Rekik, I.: Brain multiplexes reveal morphological connectional biomarkers fingerprinting late brain dementia states. Sci. Rep. **8**, 4103 (2018)
9. Wang, B., Zhu, J., Pierson, E., Ramazzotti, D., Batzoglou, S.: Visualization and analysis of single-cell RNA-seq data by kernel-based similarity learning. Nature **70**, 869–79 (2017)
10. Xu, C., Tao, D., Xu, C.: Multi-view intact space learning. IEEE Trans. Pattern Anal. Mach. Intell. **37**, 2531–2544 (2015)
11. Harville, D.A.: Maximum likelihood approaches to variance component estimation and to related problems. J. Am. Stat. Assoc. **72**, 320–338 (1977)

Neonatal Morphometric Similarity Networks Predict Atypical Brain Development Associated with Preterm Birth

Paola Galdi[1(✉)], Manuel Blesa[1], Gemma Sullivan[1], Gillian J. Lamb[1], David Q. Stoye[1], Alan J. Quigley[2], Michael J. Thrippleton[3], Mark E. Bastin[3], and James P. Boardman[1,3]

[1] MRC Centre for Reproductive Health, University of Edinburgh, Edinburgh, UK
paola.galdi@ed.ac.uk
[2] Department of Radiology, Royal Hospital for Sick Children, Edinburgh, UK
[3] Centre for Clinical Brain Sciences, University of Edinburgh, Edinburgh, UK

Abstract. Morphometric similarity networks (MSNs) have been recently proposed as a novel, robust, and biologically plausible approach to generate structural connectomes from neuroimaging data. In this work, we apply this method to multi-centre neonatal data (postmenstrual age range: 37–45 weeks) to predict brain dysmaturation in preterm infants. To achieve this goal, we combined different imaging sequences (diffusion and structural MRI) to extract a set of metrics from cortical and subcortical brain regions (e.g. regional volumes, diffusion tensor metrics, neurite orientation dispersion and density imaging features) which were used to construct a similarity network. A regression model was then trained to predict postmenstrual age at the time of scanning from inter-regional connections. Finally, to quantify brain maturation, the Relative Brain Network Maturation Index (RBNMI) was computed as the difference between predicted and actual age. The model predicted chronological age with a mean absolute error of 0.88 (± 0.63) weeks, and it consistently predicted preterm infants to have a lower RBNMI than term infants. We conclude that MSNs derived from multimodal imaging predict chronological brain development accurately, and provide a data-driven approach for defining cerebral dysmaturation associated with preterm birth.

Keywords: Morphometric similarity networks · Preterm brain
Brain developmental delay · Multi-modal MRI

P. Galdi and M. Blesa—These authors contributed equally to the work.

Electronic supplementary material The online version of this chapter (https://doi.org/10.1007/978-3-030-00755-3_6) contains supplementary material, which is available to authorized users.

G. Wu et al. (Eds.): CNI 2018, LNCS 11083, pp. 47–57, 2018.
https://doi.org/10.1007/978-3-030-00755-3_6

1   Introduction

Preterm birth is associated with a distinct brain magnetic resonance image (MRI) phenotype that includes generalized and specific dysconnectivity of neural systems and increased risk of neurocognitive and psychiatric impairment [1,2]. Different structural properties derived from MRI data, such as fractional anisotropy, mean, axial and radial diffusivity and neurite density index, have shown age related differences in the developing brain [3,4] and have been linked to altered neurodevelopment in children born preterm [5].

In recent years, several studies have used connectomics to study the developing brain and to investigate atypical development and dysmaturation [3,6,7]. The majority of connectomics work in neonates is based on single modes of data such as diffusion MRI (dMRI) tractography or resting-state functional connectivity (rsfMRI). An alternative method to model connectivity is the structural covariance network (SCN) approach [8] that works by calculating the covariance between regional measurements across subjects, resulting in a single network for the entire population. Other approaches have constructed subject-specific SCNs [9,10], but these techniques have been restricted to the use of morphometric variables available through standard structural T1-weighted MRI sequences and by using a single metric to assess the "connectivity" between nodes. To the best of our knowledge, only Shi et al. used SCNs in addition to white matter (WM) networks to provide evidence of the existence of brain alteration in neonates at genetic risk for schizophrenia [11]. The main limitation of the mentioned methods is that they do not take advantage of additional information that can be derived combining multiple modalities. In recent work Ball et al. applied canonical correlation analysis to multi-modal imaging and clinical data from preterm infants [12]. The study identified specific patterns of neonatal neuroanatomic variation that related to perinatal environmental exposures and correlated with functional outcome. They show that data-driven approaches which incorporate broad image phenotypes improve characterisation of atypical brain development after preterm birth.

In this work, we investigated whether *connectomes* derived from multimodal data within a predicting framework capture novel information about brain dysmaturation in preterm infants. Morphometric similarity networks (MSN) model the inter-regional correlations of multiple macro- and micro-structural multimodal MRI variables in a single individual. MSNs were originally devised to understand better how human cortical networks underpin individual differences in psychological functions [13]. The method works by computing a number of metrics for each region of interest (ROI) derived from different MRI sequences which are then arranged in a vector. The aim is to obtain a multidimensional description of the structural properties of the ROIs. The MSN is then built considering the ROIs as nodes and modelling connection strength as the correlation between pairs of ROI vectors. The pattern of correlations can be conceptualized as a "fingerprint" of an individual's brain. Here, the edges of individual MSNs were used to train a regression model to predict postmenstrual age (PMA) at the time of MRI acquisition. Then, to quantify brain maturation, the Relative

Brain Network Maturation Index (RBNMI) [7] was computed as the difference between predicted and actual age. We aimed to test the hypothesis that RBNMI is reduced in preterm infants at term equivalent age compared with infants born at term.

2 Methods

2.1 Participants and Data Acquisition

We combined neonatal brain MRI data collected in our institution with data from the first release of the Developing Human Connectome Project[1] (dHCP) in order to achieve balance across the age range of interest (37–45 weeks PMA). The total study sample consisted of data of 95 subjects.

The Edinburgh Birth Cohort (TEBC). Participants were recruited as part of a longitudinal study designed to investigate the effects of preterm birth on brain structure and outcome.[2] The study was conducted according to the principles of the Declaration of Helsinki, and ethical approval was obtained from the UK National Research Ethics Service. Parents provided written informed consent. A total of 55 neonates (30 preterm, 25 term, PMA range 38–45 weeks) underwent MRI at term equivalent age at the Edinburgh Imaging Facility: Royal Infirmary of Edinburgh, University of Edinburgh, Scotland, UK. A Siemens MAGNETOM Prisma 3 T MRI clinical scanner (Siemens Healthcare Erlangen, Germany) and 16-channel phased-array paediatric head coil were used to acquire: 3D T1-weighted MPRAGE (T1w) (acquired voxel size $= 1$ mm isotropic); 3D T2-weighted SPACE (T2w) (voxel size $= 1$ mm isotropic); and two dMRI acquisitions, the first one using a protocol consisting of 8 baseline volumes $(b = 0 \, \text{s/mm}^2)$ and 64 volumes with $b = 750 \, \text{s/mm}^2$, the second one consisting of 8 baseline volumes, 3 volumes with $b = 200 \, \text{s/mm}^2$, 6 volumes with $b = 500 \, \text{s/mm}^2$ and 64 volumes with $b = 2500 \, \text{s/mm}^2$. Diffusion-weighted single-shot spin-echo echo planar imaging (EPI) volumes with 2-fold simultaneous multislice and 2-fold in-plane parallel imaging acceleration were acquired with 2 mm isotropic voxels; both acquisitions had the same echo and repetition time. Infants were scanned without sedation. All images were reviewed by an expert in paediatric neuroadiology (AJQ), and none contained major focal parenchymal lesions.

Developing Human Connectome Project. The goal of the dHCP is to create a dynamic map of human brain connectivity from 20 to 45 weeks PMA. The infants were scanned using optimized protocols for structural T1w and T2w, rsfMRI, and dMRI. The first release consists of images of 40 term infants (PMA range 37–44 weeks).

---

[1] http://www.developingconnectome.org/project/.
[2] http://www.tebc.ed.ac.uk.

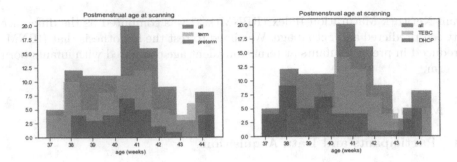

Fig. 1. Distribution of PMA at scanning in the pooled sample (37–45 weeks PMA). In the left panel, age distributions in term and preterm infants are shown. In the right panel, the age distributions in the TEBC and dHCP populations are shown.

The distribution of PMA at scan for the pooled data, each cohort, and the term and preterm groups is shown in Fig. 1. The neonates from both cohorts were separated in two groups: 65 term (PMA at scan range 37–44 weeks) and 30 preterm infants at term equivalent age (PMA at scan range 38–45 weeks). All the term infants were born after 36 week of gestation, while the preterm children were born before 32 weeks.

2.2 Data Preprocessing

Structural data were preprocessed using the dHCP minimal processing pipeline described by Makropoulos et al. [14]. The result of this processing is the parcellation of each structural image into 87 ROIs [15].

Diffusion MRI processing of TEBC dataset was performed as follows: for each subject the two dMRI acquisitions were first concatenated and then denoised using a PCA-based algorithm [16]; the eddy current, head movement and EPI geometric distortions were corrected using outlier replacement and slice-to-volume registration [17–20]; bias field inhomogeneity correction was performed by calculating the bias field of the mean b0 volume and applying the correction to all the volumes [21]. The dMRI data from the dHCP was already preprocessed [22].

For both datasets, the mean b0 volume of each subject was co-registered to their T2w volume using boundary-based registration [23], then the inverse transformation was used to propagate the ROI labels to the dMRI. For each ROI, two metrics were computed in the structural space: ROI volume and the mean T1w/T2w signal ratio. In dMRI space, seven additional metrics were also computed: the mean of each tensor-derived metric (FA: fractional anisotropy, MD: mean diffusivity, AD: axial diffusivity and RD: radial diffusivity) and the mean of the three Neurite Orientation Dispersion and Density Imaging (NODDI) parameters (ICVF: intracellular volume fraction, ODI: orientation dispersion index and VISO: isotropic volume fraction) [24]. For the tensor derived metrics,

$b = 750\,\text{s/mm}^2$ was used for the TEBC and $b = 1000\,\text{s/mm}^2$ for the dHCP data. NODDI maps were calculated with the default parameters.

2.3   Data Harmonization

We used ComBat to harmonise data because of its efficacy for removing scanner variability while preserving biological variation [25]. Each image metric was harmonized separately using gender, gestational age at birth, PMA at scan and prematurity (coded as a binary variable) as biological covariates of interest.

2.4   Network Construction

The MSN for each subject was constructed by selecting 82 of the total 87 ROIs (excluding CSF and background parcels); each of the ROI metrics was normalized (z-scored) and then Pearson correlations were computed between the vectors of metrics from each pair of ROIs. In this way, the nodes of each network are the ROIs and the edges represent the morphometric similarity between the two related ROIs (see Fig. 2 for a graphic overview of the approach). For comparison, we also computed single-metric networks: in this case connections were assigned a weight computed as the absolute difference between ROI values [10]. Note that we did not control for the effect of brain size; this is an aspect worth investigating in future work.

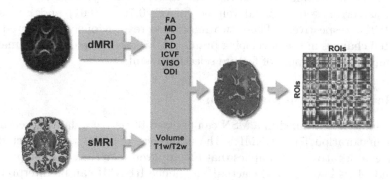

Fig. 2. Individual MSN construction. Different metrics are extracted from diffusion and structural MRI data. The same parcellation is applied to both image types and the average metric values are computed for each ROI. A connectivity matrix, representing MSN connections, is built by computing the Pearson correlation between the vectors of metrics of each pair of ROIs.

2.5   Regression Model

We trained a linear regression model with elastic net regularisation to predict PMA at scan in both preterm and term infants starting from individual

MSNs. For each subject, the edges of the MSN (inter-regional correlations) were concatenated to form a feature vector to be given as input to the regression model. Following the approach of [7], prediction performances were evaluated with a Monte Carlo cross-validation scheme. This worked as follows: in each of 100 repetitions, 50 term infants were selected at random to form the training set: the remaining 15 full term infants and 15 premature infants selected at random constituted the test set. Then, the mean absolute error (MAE) was computed across subjects and repetitions. The parameters of the regression model were selected with a nested 3-fold cross-validation loop.

2.6   Feature Selection

After the preprocessing phase, nine different metrics are available for each ROI. To study which combination of features produced better performance in the age prediction task, we implemented a sequential backward feature selection scheme. Starting from the full set of features, at each iteration we removed the feature whose subtraction caused the least increase in prediction error. The prediction error was computed with a leave-one-out cross-validation scheme on term infant data only. The best performing model, which was adopted for all subsequent analyses, was based on eight features (regional volume, T1w/T2w ratio, FA, AD, RD, ICVF, VISO, ODI) with a MAE of 0.64 (±0.44) weeks. None of the single-metric networks (Sect. 2.4) outperformed the selected MSN, however it is interesting to note that the metrics that yielded the least and second least error were regional volume (MAE: 0.76 ± 0.61) and FA (MAE: 0.79 ± 0.91), respectively. These two metrics were previously associated with age-related changes in the developing brain [3, 4, 15]. See supplementary material for a comprehensive report of feature selection results.

2.7   Measuring Brain Maturation

The information contained in a MSN can be used to derive a data-driven metric of brain maturation [7]. RBNMI is the difference between the predicted and actual age: a negative score implies that the apparent age (i.e., the age predicted by the model) is lower than the actual age; hence RBNMI can be interpreted as an index of dysmaturation.

3   Results and Discussion

3.1   Data Harmonisation

We tested the efficacy of *ComBat* harmonization by comparing MSNs built before and after harmonizing the data. We used principal component analysis to perform an unsupervised dimension reduction and projected individual MSN onto the first two principal components (PCs), explaining 28% of the variance. Figure 3 shows that without harmonization points are distinctly clustered by dataset, while after harmonization there is no clear separation.

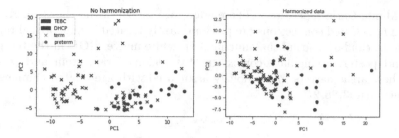

Fig. 3. Individual MSNs projected onto the first two principal components (PC1 and PC2) that explain most of the variation in the data, before harmonization (left panel) and after (right panel).

3.2 Morphometric Similarity Networks

The selected model predicted PMA at scan with a MAE of 0.88 ($\pm$0.63) weeks when tested on all data. We then computed the mean RBNMI values for each group of subjects, averaged across the 100 Monte Carlo repetitions. The group RBNMI was more negative for the preterm group (Fig. 4). The difference between the two groups was confirmed with a two-tailed t-test ($p < 0.001$) after testing for departure from normality of the RBNMI distributions with a D'Agostino and Pearson's test ($p = 0.42$ for the pretermn group; $p = 0.98$ for the term group). These results suggest that the information contained in a MSN is sufficient to detect a dysmaturation in preterm infants at term equivalent age. One interesting aspect to consider is that the MSN was built by combining features from different imaging sequences that describe complementary aspects of brain structure and have been previously studied in isolation [3,15]. Hence, to fully describe the developing brain it is crucial to follow a holistic approach, integrating information from multiple sources.

To study which connections contributed the most to age prediction, we selected only edges which were assigned a non-zero coefficient in at least 99%

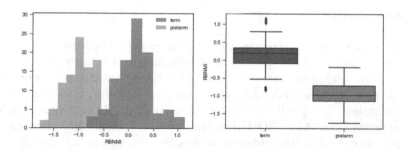

Fig. 4. Distribution of per group mean RBNMI values, averaged across test subjects within each of the 100 rounds of cross-validation, shown as histograms (left) and as box plots (right). Averaged RBNMI values in the preterm group were significantly more negative than in the term group.

of the Monte Carlo repetitions. These edges are shown in the chord diagram in Fig. 5. The selected connections are predominantly located in subcortical regions (thalamus, caudate and subthalamic nuclei); white matter ROIs in the temporal lobe and posterior cingulate cortex; frontal regions, brain stem and cerebellum. These areas have been previously associated with age-related changes and preterm birth [3, 26, 27].

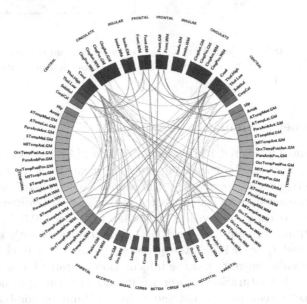

Fig. 5. Chord diagram showing MSN edges used for prediction in at least 99% of regression models in the Monte Carlo repetitions. The thickness of connections reflects the strength of the correlation between edge weights and age across subjects (positive correlations in shades of gray; negative correlations in shades of red). (Color figure online)

4   Conclusions

We used inter-regional morphometric similarities to train a predictive model for PMA at scan and to derive a data-driven metric of brain maturation, the RBNMI. We found that preterm infants at term equivalent age were consistently assigned a lower RBNMI than infants born at term, indicating delayed brain maturation. The predictive model for age performed best when structural, diffusion tensor-derived and NODDI metrics were combined, which demonstrates the importance of integrating different biomarkers to generate a global picture of the developing human brain. The motivation for using a network-based approach is indeed obtaining a whole-brain description able to capture a developmental pattern. Our results suggest that the information contained in MSNs is sufficient to train a machine learning model in the age prediction task. A second reason for working with similarities instead of single regional metrics is methodological: computing edge weights as inter-regional similarities enables an integrated representation of all available metrics in a single network; to work with the original features directly would mean either working with several networks (thus requiring a further step to integrate them) or concatenating all the features in a single predictive model, aggravating the problems related with the "curse of dimensionality". In future work, we plan to extend the presented approach to other clinical variables beyond PMA, such as clinical risk indices and neurocognitive outcome data.

Acknowledgements. We are grateful to the families who consented to take part in the study. This work was supported by Theirworld (www.theirworld.org) and was undertaken in the MRC Centre for Reproductive Health, which is funded by MRC Centre Grant (MRC G1002033). MJT was supported by NHS Lothian Research and Development Office. Participants were scanned in the University of Edinburgh Imaging Research MRI Facility at the Royal Infirmary of Edinburgh which was established with funding from The Wellcome Trust, Dunhill Medical Trust, Edinburgh and Lothians Research Foundation, Theirworld, The Muir Maxwell Trust and many other sources; we thank the University's Imaging Research staff for providing the infant scanning. Part of the results were obtained using data made available from the Developing Human Connectome Project (King's College London - Imperial College London - Oxford University dHCP Consortium) funded by the European Research Council under the European Union's Seventh Framework Programme (FP/2007–2013)/ERC Grant Agreement no. 319456.

References

1. Batalle, D., Edwards, A.D., O'Muircheartaigh, J.: Annual research review: not just a small adult brain: understanding later neurodevelopment through imaging the neonatal brain. J. Child Psychol. Psychiatr. **59**(4), 350–371 (2018)
2. Telford, E.J., Cox, S.R., Fletcher-Watson, S., Anblagan, D., Sparrow, S., et al.: A latent measure explains substantial variance in white matter microstructure across the newborn human brain. Brain Struct. Funct. **222**(9), 4023–4033 (2017)
3. Batalle, D., Hughes, E.J., Zhang, H., Tournier, J.D., Tusor, N., et al.: Early development of structural networks and the impact of prematurity on brain connectivity. NeuroImage **149**, 379–392 (2017)
4. Batalle, D., O'Muircheartaigh, J., Makropoulos, A., Kelly, C.J., Dimitrova, R., et al.: Different patterns of cortical maturation before and after 38 weeks gestational age demonstrated by diffusion MRI in vivo. NeuroImage (2018). https://doi.org/10.1016/j.neuroimage.2018.05.046
5. Counsell, S.J., Ball, G., Edwards, A.D.: New imaging approaches to evaluate newborn brain injury and their role in predicting developmental disorders. Cur. Opin. Neurol. **27**(2), 168–175 (2014)
6. Van Den Heuvel, M.P., Kersbergen, K.J., De Reus, M.A., Keunen, K., et al.: The neonatal connectome during preterm brain development. Cereb. Cortex **25**(9), 3000–3013 (2015)

7. Brown, C.J., et al.: Prediction of brain network age and factors of delayed maturation in very preterm infants. In: Descoteaux, M., Maier-Hein, L., Franz, A., Jannin, P., Collins, D.L., Duchesne, S. (eds.) MICCAI 2017. LNCS, vol. 10433, pp. 84–91. Springer, Cham (2017). https://doi.org/10.1007/978-3-319-66182-7_10

8. Alexander-Bloch, A., Giedd, J.N., Bullmore, E.: Imaging structural co-variance between human brain regions. Nat. Rev. Neurosci. **14**(5), 322–336 (2013)

9. Li, W., et al.: Construction of individual morphological brain networks with multiple morphometric features. Front. Neuroanat. **11**, 34 (2017)

10. Mahjoub, I., Mahjoub, M.A., Rekik, I.: Brain multiplexes reveal morphological connectional biomarkers fingerprinting late brain dementia states. Sci. Rep. **8**(1), 4103 (2018)

11. Shi, F., Yap, P.T., Gao, W., Lin, W., Gilmore, J.H., Shen, D.: Altered structural connectivity in neonates at genetic risk for schizophrenia: a combined study using morphological and white matter networks. NeuroImage **62**(3), 1622–1633 (2012)

12. Ball, G., Aljabar, P., Nongena, P., Kennea, N., Gonzalez-Cinca, N., et al.: Multimodal image analysis of clinical influences on preterm brain development. Ann. Neurol. **82**(2), 233–246 (2017)

13. Seidlitz, J., Váša, F., Shinn, M., Romero-Garcia, R., Whitaker, K.J.: Morphometric similarity networks detect microscale cortical organization and predict inter-individual cognitive variation. Neuron **97**(1), 231–247.e7 (2018)

14. Makropoulos, A., Robinson, E.C., Schuh, A., Wright, R., Fitzgibbon, S., et al.: The developing human connectome project: a minimal processing pipeline for neonatal cortical surface reconstruction. NeuroImage **173**, 88–112 (2018)

15. Makropoulos, A., Aljabar, P., Wright, R., Hüning, B., Merchant, N., et al.: Regional growth and atlasing of the developing human brain. NeuroImage **125**, 456–478 (2016)

16. Veraart, J., Novikov, D.S., Christiaens, D., Ades-aron, B., Sijbers, J., Fieremans, E.: Denoising of diffusion MRI using random matrix theory. NeuroImage **142**, 394–406 (2016)

17. Andersson, J.L., Skare, S., Ashburner, J.: How to correct susceptibility distortions in spin-echo echo-planar images: application to diffusion tensor imaging. NeuroImage **20**(2), 870–888 (2003)

18. Andersson, J.L., Sotiropoulos, S.N.: An integrated approach to correction for off-resonance effects and subject movement in diffusion MR imaging. NeuroImage **125**, 1063–1078 (2016)

19. Andersson, J.L., Graham, M.S., Zsoldos, E., Sotiropoulos, S.N.: Incorporating outlier detection and replacement into a non-parametric framework for movement and distortion correction of diffusion MR images. NeuroImage **141**, 556–572 (2016)

20. Andersson, J.L., Graham, M.S., Drobnjak, I., Zhang, H., Filippini, N., Bastiani, M.: Towards a comprehensive framework for movement and distortion correction of diffusion MR images: Within volume movement. NeuroImage **152**, 450–466 (2017)

21. Tustison, N.J., Avants, B.B., Cook, P.A., Zheng, Y., Egan, A., et al.: N4ITK: improved N3 bias correction. IEEE Trans. Med. Imaging **29**(6), 1310–1320 (2010)

22. Bastiani, M., Andersson, J., Cordero-Grande, L., Murgasova, M., Hutter, J., et al.: Automated processing pipeline for neonatal diffusion MRI in the developing human connectome project. NeuroImage (2018). https://doi.org/10.1016/j.neuroimage.2018.05.064

23. Greve, D.N., Fischl, B.: Accurate and robust brain image alignment using boundary-based registration. NeuroImage **48**(1), 63–72 (2009)

24. Zhang, H., Schneider, T., Wheeler-Kingshott, C.A., Alexander, D.C.: NODDI: practical in vivo neurite orientation dispersion and density imaging of the human brain. NeuroImage **61**(4), 1000–1016 (2012)
25. Fortin, J.P., Parker, D., Tunç, B., Watanabe, T., Elliott, M.A., et al.: Harmonization of multi-site diffusion tensor imaging data. NeuroImage **161**, 149–170 (2017)
26. Boardman, J.P., Counsell, S.J., Rueckert, D., Kapellou, O., Bhatia, K.K., et al.: Abnormal deep grey matter development following preterm birth detected using deformation-based morphometry. NeuroImage **32**(1), 70–78 (2006)
27. Ball, G., Boardman, J.P., Aljabar, P., Pandit, A., Arichi, T., et al.: The influence of preterm birth on the developing thalamocortical connectome. Cortex **49**(6), 1711–1721 (2013)

Heritability Estimation of Reliable Connectomic Features

Linhui Xie[1], Enrico Amico[6,7], Paul Salama[1], Yu-chien Wu[2],
Shiaofen Fang[4], Olaf Sporns[5], Andrew J. Saykin[2], Joaquín Goñi[6,7],
Jingwen Yan[2,3(✉)], and Li Shen[8(✉)]

[1] Department of Electrical and Computer Engineering, Indiana University-Purdue
University Indianapolis, Indianapolis, IN, USA
[2] Department of Radiology and Imaging Sciences,
Indiana University School of Medicine, Indianapolis, IN, USA
[3] Department of BioHealth Informatics, Indiana University-Purdue University
Indianapolis, Indianapolis, IN, USA
jingyan@iupui.edu
[4] Department of Computer Science, Indiana University-Purdue
University Indianapolis, Indianapolis, IN, USA
[5] Department of Psychological and Brain Science,
Indiana University, Bloomington, IN, USA
[6] School of Industrial Engineering, Purdue University, West Lafayette, IN, USA
[7] Weldon School of Biomedical Engineering, Purdue University,
West Lafayette, IN, USA
[8] Biostatistics, Epidemiology and Informatics, University of Pennsylvania,
Philadelphia, PA, USA
li.shen@pennmedicine.upenn.eduu

Abstract. Brain imaging genetics is an emerging research field to explore the underlying genetic architecture of brain structure and function measured by different imaging modalities. However, not all the changes in the brain are a consequential result of genetic effect and it is usually unknown which imaging phenotypes are promising for genetic analyses. In this paper, we focus on identifying highly heritable measures of structural brain networks derived from diffusion weighted imaging data. Using the twin data from the Human Connectome Project (HCP), we evaluated the reliability of fractional anisotropy measure, fiber length and fiber number of each edge in the structural connectome and seven network level measures using intraclass correlation coefficients. We then estimated the heritability of those reliable network measures using SOLAR-Eclipse software. Across all 64,620 network edges between 360 brain regions in the Glasser parcellation, we observed ∼5% of them with significantly high heritability in fractional anisotropy, fiber length or fiber number. All the tested network level measures, capturing the network integrality, segregation or resilience, are highly heritable, with variance explained by the additive genetic effect ranging from 59% to 77%.

Keywords: Structural connectivity · Heritability · Reliability · HCP

© Springer Nature Switzerland AG 2018
G. Wu et al. (Eds.): CNI 2018, LNCS 11083, pp. 58–66, 2018.
https://doi.org/10.1007/978-3-030-00755-3_7

1   Introduction

Brain imaging genetics is an emerging research field that integrates genotyping and neuroimaging data to explore the underlying genetic architecture of brain structure and function. Genetic analysis of imaging measures not only allows the detection of risk variants associated with diseases, but also provides insights into the underlying biological mechanism of preclinical brain changes. However, not all the changes in the brain are a consequential result of genetic effect. It is usually unknown which imaging phenotypes are promising for genetic analyses. Therefore, prior to that, it is important to quantify the degree to which brain imaging phenotypes can be attributed to genetic effect using heritability analysis.

Recently, substantial attention has been drawn to the genetic influence on structural brain connectivity, which appeared to be altered in heritable diseases (e.g. Alzheimer's disease [13]). One widely analyzed measure is fractional anisotropy (FA), an measure of fiber integrity very sensitive to the white matter changes in various diseases [9]. Brain-wide, regional and voxel level FA measures have all been found to be highly and significantly heritable [1,9]. Other features that have been investigated include white matter fiber tract shapes [8], white matter volume, network level characteristic path length and clustering coefficient [1], and fiber orientation distribution [15]. However, these studies mostly focus on the heritability of tracts (i.e. white matter ROIs) themselves, but not the resulting anatomical connections of the human brain (i.e. connectome). To this end, the heritability of brain connectomic features remains largely unknown.

To bridge this gap, we propose to perform a comprehensive heritability analysis of anatomical brain networks using the twin data from the Human Connectome Project (HCP) [19]. We employ a new brain parcellation defined based on functional MRI (fMRI) to generate brain networks with improved anatomical precision, enabling us to examine the genetic influence on the structural coordination within/between functional brain circuits. With three sessions of diffusion weighted imaging (DWI) scans for each individual, we first evaluate the reliability of three edge-level measures, including fractional anisotropy, fiber length and fiber number, and seven network-level measures using intraclass correlation coefficients (ICC). The heritability of those reliable network measures were then estimated using Sequential Oligogenic Linkage Analysis Routines (SOLAR)-Eclipse software. Across all 64,620 edges between 360 ROIs, ~5% of them show significantly high heritability in fractional anisotropy, fiber length or fiber number. Top functional brain circuits connected by these heritable edges include visual and default mode network (DMN). All the tested network level measures, capturing the integrity, segregation or resilience of brain networks, are highly heritable with variance explained by the additive genetic effect ranging from 59% to 77%.

2   Method

HCP Data. We downloaded high spatial resolution DWI data from the Human Connectome Project (HCP) [19]. In total, there are 179 pair of twins

(Age: 29.1 ± 3.68), including 136 mono-zygotic females, 98 mono-zygotic males, 68 di-zygotic females and 56 di-zygotic males. DWI data was processed following the MRtrix3 guidelines [18]. More specifically, we first generated a tissue-segmented image appropriate for anatomically constrained tractography [16]. Then, we estimated the multi-shell multi-tissue response function [3] and performed the multi-shell, multi-tissue constrained spherical deconvolution [7]. Afterwards, we generated the initial tractogram with 10 million streamlines (maximum tract length = 250, FA cutoff = 0.06) and applied the successor of Spherical-deconvolution Informed Filtering of Tractograms (SIFT2) methodology [17]. Compared to SIFT, SIFT2 generates more biologically accurate measures of fiber connectivity whilst making use of the complete streamlines reconstruction [17]. Finally, we mapped the SIFT2 output streamlines onto the Glasser atlas with 360 ROIs [5] to produce the structural connectome. The final brain networks were constructed using fibers going through white matter and connecting Glasser ROIs. In this project, we focused on three edge-level measures, including fractional anisotropy (FA), length of fibers (LOF) and number of the fibers (NOF). In addition, we binarized the brain network and calculated seven network-level topological features characterizing the integrity, segregation and resilience of brain networks [14] (see Table 1).

Reliability of Connectomic Features. Tractography-based networks is known to have an issue on measurement reliability. To investigate the precision of connectomic features, we estimated the test-retest reliability by comparing three DWI data sets of the same individuals acquired at different time points. We calculated intraclass correlation coefficients (ICC) for each brain connectomic feature to evaluate their reliability [12]. All connectomic features with good/excellent reliability (ICC ≥ 0.75) are included for the subsequent heritability analysis [10].

Heritability Analysis. Heritability is defined as the proportion of phenotypic variance attributable to genetic effect. In this project, we estimated the

Table 1. Heritability of topological features derived from brain networks.

Topological features	ICC	h^2	Std. error	P-value	Variance (covariates)
Assortativity coefficient	0.92	0.59	0.06	3.50×10^{-13}	0.04
Local efficiency	0.89	0.76	0.04	1.36×10^{-24}	0.18
Modularity	0.87	0.70	0.05	3.02×10^{-19}	0.11
Transitivity	0.89	0.77	0.04	3.90×10^{-24}	0.16
Cluster coefficient	0.89	0.76	0.04	1.37×10^{-24}	0.17
Global efficiency	0.87	0.75	0.04	4.88×10^{-23}	0.16
Characteristic path length	0.85	0.72	0.04	5.71×10^{-23}	0.02

heritability of brain connectomoic features extracted from twin subjects in the HCP cohort without using any genetic data. SOLAR-Eclipse software tool is chosen over traditional ACE modal due to its capability in evaluating the covariate effects, significance of heritability and standard error for each trait [4,9]. It requires three inputs: phenotype traits, covariates measures and a kinship matrix indicating the pairwise relationship between twin individuals. A maximum likelihood variance decomposition method is applied to estimate the variance explained by additive genetic factors and environmental factors respectively. The output from SOLAR-Eclipse includes heritability (h^2), standard error and the corresponding significance p-value for each feature. We estimated the heritability of all brain connectomic features, including FA, LOF, NOF of 64,620 edges and 7 network level measures (Table 1). Network level measures are derived from binarized brain network in a way that the weight of the link is set to one when it exists and zero otherwise [14]. Prior to the heritability analysis, inverse Gaussian transformation was applied to ensure normality of all the measures. Since many previous studies have reported the effect of age (linear/nonlinear), gender and their interactions on structural brain connectivity [2,6,11,22], all heritability analyses were conducted with age at scan, age^2, sex, age × sex and age^2 × sex as covariates. In addition, we extracted the total variance explained by all covariate variables.

3   Results

Reliability of Brain Connectomic Features. Shown in Fig. 1(a) is the scatter plot of edge-level reliability against heritability estimated in SOLAR-Eclipse. Each dot corresponds to one edge and the color indicates the significance of the heritability. For FA, LOF and NOF measures, 11.13%, 9.95% and 45.54% of total edges respectively show consistency across three sessions with good/excellent reliability score (ICC $\geq$ 0.75). In total, 43,051 out of 193,860 edge-level features passed the reliability test and their heritability patterns will be further analyzed. All tested network level measures show very good reproducibility across sessions, with the ICC value ranging from 0.85 to 0.92 (see Table 1). Since we focus on the features reproducible across three sessions, the heritability analysis was only performed on the DWI data from one session.

Heritability of Edge-Level Measures. After excluding the edges without passing reliability test, there are 5105 edges whose FA show significantly high heritability after stringent Bonferroni correction (p $\leq$ 0.05/(64,620×3) = 2.58e−7). For LOF and NOF measures, there are 2687 edges and 7311 edges passing the significance threshold respectively. From Fig. 1(b), we observe that the heritability (h^2) of FA measure is between 0.4 and 0.85. LOF and NOF measures show similar heritability distribution, but there is much less edges with very high heritability ($h^2 \geq 0.8$).

Shown in Fig. 1(c) are the heatmaps of anatomical connection matrix with ICC $\geq$ 0.75 and p $\leq$ 2.58e − 7 for FA, LOF and NOF measures respectively.

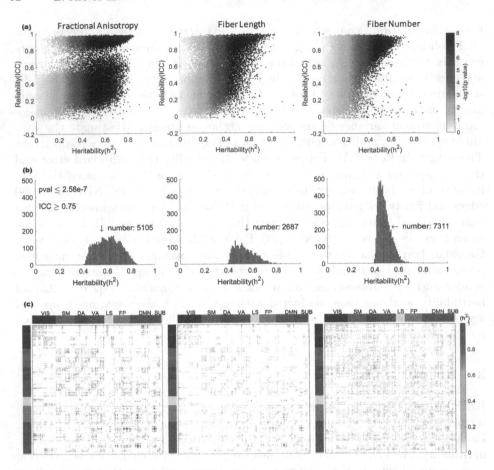

Fig. 1. Heritability distribution of all significant and reliable edges. (a) Scatter plots of reliability against heritability. Dot color indicates log-transformed p-value. (b) Histogram for reliabe edges. (c) Heatmap of anatomical connection matrix. Rows and columns are reordered to form seven functional groups corresponding to Yeo parcellation. Top and side color panels indicate the corresponding Yeo parcellation of each ROI. The last subcortical (SUB) group is added to complement the Yeo atlas. (Color figure online)

Glasser brain regions were reordered to form seven functional groups defined in Yeo parcellation (Fig. 2) [21]. Subcortical part was added to complement Yeo atlas. For all three edge-level measures, the majority of those significantly heritable and reliable edges are located within default mode network, within visual circuit, or connecting default mode network with other circuits, such as Ventral Attention and Frontal-Parietal. Edges connecting Visual and Somato-Motor circuits show the highest average heritability ($h^2 = 0.69$) in FA measure. For LOF and NOF, the edges with the highest average heritability are from Limbic system ($h^2 = 0.64$ for LOF and $h^2 = 0.49$ for NOF).

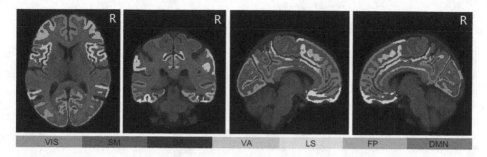

Fig. 2. Brain map of Yeo parcellation in MNI space. From left to right: axial view, coronal view, sagittal view (Left) and sagittal view (Right). The bottom color panel indicates the color scheme of different regions: Visual (VIS), Somato-Motor (SM), Dorsal Attention (DA), Ventral Attention (VA), Limbic system (LS), Fronto-Parietal (FP) and Default Mode Network (DMN). (Color figure online)

For each type of measures, we further ranked the edges based on their heritability (h^2) and examined the brain regions connected by those top heritable edges. In Fig. 3(a) are the heatmaps showing the heritability of top 0.5% edges in FA, LOF and NOF respectively. In the brain connectivity map (Fig. 3(b)), we observed that many top heritable edges are within the frontal lobe. several brain regions in occipital lobe (primary visual cortex) as hubs connected by some highly heritable edges for FA and LOF. These fiber tracts belong to white matter region inferior longitudinal fasciculus, whose regional FA value is previously identified to be highly heritable [9]. In addition, we found that the length of Cingulum tracts (vertical lines in the middle of the brain) are also largely controlled by the genetic factors, with h^2 around 0.65. Its FA measure was also previously reported to be heritable with $h^2 = 0.81$ [9]. For NOF, top heritable edges show a different spatial pattern and are more evenly distributed across the whole brain. Functional brain circuits that are mostly connected by these top heritable edges are DMN and FP (Fig. 3(c)). Finally, we examined the expression patterns of those brain regions involved in the DMN circuit in Allen Human Brain Atlas, including medial prefrontal cortex, angular, precuneus, posterior cingulate cortex and hippocampus. Interestingly, many of these brain regions connected by heritable edges show very similar gene expression patterns.

Heritability of Network-Level Measures. For each individual, we extracted seven network-level topological features to evaluate the integrity, segregation and resilience of brain network, including assortativity coefficient, modularity, local efficiency, cluster efficiency, transitivity, characteristic path length and its inverse measure global efficiency. The detailed description of these measures is available in [14]. Shown in Table 1 is the summary of estimated heritability for all topological features. Five covariates can explain ∼15% variance of all topological features except for assortativity coefficient and characteristic path length. Sex and age^2 × sex are the only factors that exhibit significant influence

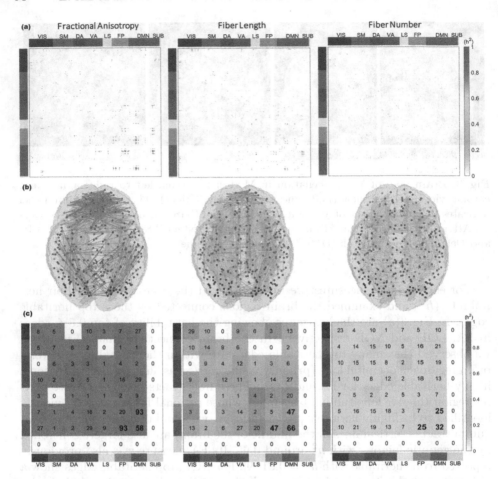

Fig. 3. Heritability of top 0.5% edges ranked by h^2. (a) Heatmap of anatomical connection matrix. (b) Heritability of edge-level measures in the brain map. Node color indicates different Yeo functional groups. (c) Heatmap showing total number and average h^2 value of edges connecting each pair of functional groups in Yeo parcellation. Top and side color panels indicate the corresponding Yeo parcellation of each ROI. The last subcortical (SUB) group is added to complement the Yeo atlas. (Color figure online)

on the network topology heritability. Assortativity coefficient has the highest reliability, but only 58% of variance can be attributed to the additive genetic effect. The other six features are estimated to have similar heritability around 0.75. These findings are consistent with previous studies, e.g. characteristic path length and clustering coefficient of anatomical brain network are highly heritable [1]. Our heritability estimation is slightly higher than theirs, possibly due to the selection of different brain parcellation schemes.

4 Conclusion

We performed a comprehensive heritability analysis for both edge-level and network-level brain connectomic features. Unlike previous studies that largely focus on the tracts (i.e. white matter ROIs), we used a new brain parcellation to construct brain networks (connectome) with improved anatomical precision. Our results show the degree to which the genetic factors may influence the structural coordination between/within functional brain circuits. Many edges/tracts were found to be highly heritable, particularly those connecting default mode network circuit or visual circuit. This is consistent with another finding that hub regions in the default mode network have very similar gene expression patterns [20]. All seven tested network-level features are reliable and significantly heritable. Future effort is warranted to investigate the genetic variations underlying these heritable connectomic features.

References

1. Bohlken, M.M., et al.: Heritability of structural brain network topology: a dti study of 156 twins. Hum. Brain Mapp. **35**(10), 5295–305 (2014). https://doi.org/10.1002/hbm.22550
2. Burzynska, A.Z., et al.: Age-related differences in white matter microstructure: region-specific patterns of diffusivity. Neuroimage **49**(3), 2104–2112 (2010)
3. Christiaens, D., Reisert, M., Dhollander, T., Sunaert, S., Suetens, P., Maes, F.: Global tractography of multi-shell diffusion-weighted imaging data using a multi-tissue model. Neuroimage **123**, 89–101 (2015)
4. Ganjgahi, H., Winkler, A.M., Glahn, D.C., Blangero, J., Kochunov, P., Nichols, T.E.: Fast and powerful heritability inference for family-based neuroimaging studies. NeuroImage **115**, 256–268 (2015)
5. Glasser, M.F., et al.: A multi-modal parcellation of human cerebral cortex. Nature **536**(7615), 171–178 (2016)
6. Gong, G., He, Y., Evans, A.C.: Brain connectivity: gender makes a difference. Neuroscientist **17**(5), 575–591 (2011)
7. Jeurissen, B., Tournier, J.D., Dhollander, T., Connelly, A., Sijbers, J.: Multi-tissue constrained spherical deconvolution for improved analysis of multi-shell diffusion MRI data. NeuroImage **103**, 411–426 (2014)
8. Jin, Y., et al.: Heritability of white matter fiber tract shapes: a HARDI study of 198 twins. In: Liu, T., Shen, D., Ibanez, L., Tao, X. (eds.) MBIA 2011. LNCS, vol. 7012, pp. 35–43. Springer, Heidelberg (2011). https://doi.org/10.1007/978-3-642-24446-9_5
9. Kochunov, P., et al.: Heritability of fractional anisotropy in human white matter: a comparison of Human Connectome Project and ENIGMA-DTI data. Neuroimage **111**, 300–11 (2015). https://doi.org/10.1016/j.neuroimage.2015.02.050
10. Koo, T.K., Li, M.Y.: A guideline of selecting and reporting intraclass correlation coefficients for reliability research. J. Chiropr. Med. **15**(2), 155–163 (2016)
11. Lopez-Larson, M.P., Anderson, J.S., Ferguson, M.A., Yurgelun-Todd, D.: Local brain connectivity and associations with gender and age. Dev. Cogn. Neurosci. **1**(2), 187–197 (2011)

12. McGraw, K.O., Wong, S.P.: Forming inferences about some intraclass correlation coefficients. Psychol. Methods **1**(1), 30 (1996)
13. Nir, T.M., et al.: Effectiveness of regional DTI measures in distinguishing Alzheimer's disease, MCI, and normal aging. NeuroImage Clin. **3**, 180–195 (2013)
14. Rubinov, M., Sporns, O.: Complex network measures of brain connectivity: uses and interpretations. Neuroimage **52**(3), 1059–1069 (2010)
15. Shen, K.K., et al.: Investigating brain connectivity heritability in a twin study using diffusion imaging data. Neuroimage **100**, 628–41 (2014). https://doi.org/10.1016/j.neuroimage.2014.06.041
16. Smith, R.E., Tournier, J.D., Calamante, F., Connelly, A.: Anatomically-constrained tractography: improved diffusion MRI streamlines tractography through effective use of anatomical information. Neuroimage **62**(3), 1924–1938 (2012)
17. Smith, R.E., Tournier, J.D., Calamante, F., Connelly, A.: SIFT2: enabling dense quantitative assessment of brain white matter connectivity using streamlines tractography. Neuroimage **119**, 338–351 (2015)
18. Tournier, J., Calamante, F., Connelly, A., et al.: MRtrix: diffusion tractography in crossing fiber regions. Int. J. Imaging Syst. Technol. **22**(1), 53–66 (2012)
19. Van Essen, D.C., et al.: The WU-Minn human connectome project: an overview. Neuroimage **80**, 62–79 (2013)
20. Wang, G.Z., et al.: Correspondence between resting-state activity and brain gene expression. Neuron **88**(4), 659–666 (2015)
21. Yeo, B.T., et al.: The organization of the human cerebral cortex estimated by intrinsic functional connectivity. J. Neurophysiol. **106**(3), 1125–1165 (2011)
22. Zhao, T., et al.: Age-related changes in the topological organization of the white matter structural connectome across the human lifespan. Hum. Brain Mapp. **36**(10), 3777–3792 (2015)

Topological Data Analysis of Functional MRI Connectivity in Time and Space Domains

Keri L. Anderson[✉], Jeffrey S. Anderson, Sourabh Palande, and Bei Wang

University of Utah, Salt Lake City, USA
andersonkeril@gmail.com

Abstract. The functional architecture of the brain can be described as a dynamical system where components interact in flexible ways, constrained by physical connections between regions. Using correlation, either in time or in space, as an abstraction of functional connectivity, we analyzed resting state fMRI data from 1003 subjects. We compared several data preprocessing strategies and found that independent component-based nuisance regression outperformed other strategies, with the poorest reproducibility in strategies that include global signal regression. We also found that temporal vs. spatial functional connectivity can encode different aspects of cognition and personality. Topological analyses using persistent homology show that persistence barcodes are significantly correlated to individual differences in cognition and personality, with high reproducibility. Topological data analyses, including approaches to model connectivity in the time domain, are promising tools for representing high-level aspects of cognition, development, and neuropathology.

1   Introduction

Recent advances in the science of complex networks have utilized tools from topological data analysis, in particular, persistent homology [10,29]. This approach investigates connections between constituent parts of networks using powerful and flexible algorithms designed to encode and measure the persistence of relationships across multiple scales. As a discipline, topological data analysis combines algebraic topology and other tools from pure and applied mathematics to support a widening array of applications for studying the architecture of dynamic and complex networks. Although graph-theoretic connectivity, a relatively simple type of topological analysis, has found wide application in functional neuroimaging [23], more advanced uses of topological data analysis are emerging.

Electronic supplementary material The online version of this chapter (https://doi.org/10.1007/978-3-030-00755-3_8) contains supplementary material, which is available to authorized users.

We evaluate the performance of graph-theoretic and persistent homology metrics on both time and space formulations of functional connectivity and use the Human Connectome Project dataset [26] to evaluate how the topological architecture of brain networks is related to cognitive function. Since image processing strategies traditionally used for resting state fMRI data have been optimized for conventional types of analysis, we also evaluate the effects of different preprocessing pipelines. Specifically, we measure the reproducibility of graph-theoretic metrics obtained from both time and space dimensions, following several preprocessing pipelines to assess the potential impact of processing strategies on the ability of graph-theoretic metrics to identify individual differences in connectivity and related cognition, behavior and personality.

Our analyses demonstrate that: (1) Graph-theoretic and topological analyses performed on connectivity across time and space are correlated with distinct aspects of cognitive function. (2) The barcode obtained from the persistent homology of resting state fMRI data is a compact representation of information about individual differences in cognition and personality, with nearly all cognitive metrics tested showing significant correlation to topological features within brain regions known to be substrates for related cognitive functions. (3) Topological metrics show excellent reproducibility when applied to long-duration functional connectivity metrics (1 h resting state fMRI per subject). Reproducibility was highest for FIX ICA processed data [15] and consistently poorer for preprocessing strategies that include global signal regression [20].

2   Related Work and Technical Background

Dynamic Functional Connectivity. Methods for dynamical analysis of the temporal information in fMRI image data have shown promise for extracting information not available from conventional functional connectivity approaches [18]. Several of these methods involve sliding window approaches [1], where synchronization between brain regions is estimated from short epochs of time. Windowed analyses, however, obtain estimates of connectivity from short time series, resulting in limited accuracy of individual measurements.

An additional approach to component analysis includes identifying point-process temporal co-activation patterns [5,17] by clustering time points that exhibit similar activation across the brain and examining temporal structure within the relative sequence of activation of these co-activation patterns. Point-process methods such as temporal co-activation patterns have been analyzed by clustering timepoints together with similar patterns of relative activation across the brain into composite spatial nodes or networks that show hierarchical similarity structure to each other with architecture that combines features of more familiar intrinsic connectivity networks [5,17]. Nevertheless, there is a symmetry where an fMRI time series can arbitrarily be seen as connectivity between time points across the brain, where the individual time points become the nodes in a graph, with edges reflected by spatial similarity across the brain.

Topological Data Analysis (TDA) of Networks. TDA [3,13] of networks goes beyond graph-theoretic analysis by utilizing tools from computational topology to describe the architecture of networks or data structures in more flexible ways. In particular, it encodes higher order (not just pairwise) interactions in the system and studies topological features of the brain network across all possible thresholds. Persistent homology, a main ingredient in topological data analysis, is an emerging tool in studying complex networks, including, for instance, collaboration [4] and brain networks [19]. Topological methods have also shown promise in modeling transitions between brain states in functional imaging data using combined information in space and time [24].

Persistent Homology and Barcodes. Persistent homology studies the topological features of a point cloud at multiple scales; see [10] for seminal work on the topic and [3,13] for excellent surveys. We follow the illustrative example of persistent homology and metric space mapping in [27].

As illustrated in Fig. 1, we begin with a point cloud P, equipped with a (Euclidean) distance metric. For some $t \geq 0$, the union of balls of radius t, centered at the points of P, forms a topological space. As the radius t increases, we get a nested sequence of spaces referred to as a *filtration*. The radius t, which parametrizes the spaces in such filtration, is often viewed as time. Using persistent homology, we investigate the evolution in time of the topological features of spaces in the filtration.

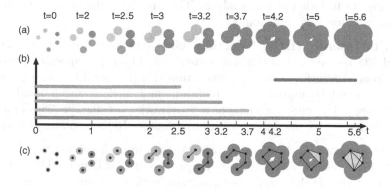

Fig. 1. Computing the persistent homology of a point cloud (image adapted from [27], Fig. 1). (Color figure online)

As t increases, we focus on the important events when the topology of the space changes. This change occurs, for example, when components merge with one another to form larger components or tunnels. We track the birth and death times of each topological feature (a component or a tunnel). The lifetime of a feature in the filtration is called its *persistence*. In Fig. 1(a), at time $t = 0$, each colored point is *born* (appears) as an independent (connected) component. At $t = 2.5$, the green component merges into the red component and *dies* (disappears).

Therefore, the green component has a persistence of 2.5. At $t = 3$, the orange component merges into the pink component and dies. Hence, it has a persistence of 3. Similarly, the blue component dies at $t = 3.2$ and the pink component dies at $t = 3.7$. At time $t = 4.2$, the collection of components forms a tunnel which has a persistence of 1.4 and disappears at $t = 5.6$. The red component born at time 0 never dies and thus it has a persistence of ∞. In Fig. 1(b), we visualize the appearance (birth), the disappearance (death) and the persistence of these topological features in the filtration via the *barcode* [13], where each feature is summarized by a horizontal bar that begins at its birth time and ends at its death time. Computationally, the above nested sequence of spaces can be combinatorially represented by a nested sequence of simplicial complexes (i.e., collections of vertices, edges and triangles) with a much smaller footprint, as illustrated in Fig. 1(c); see [9] for computational details.

3   Methods

Data Sources. Resting state fMRI data from 1003 participants (534 female, mean age = 29.45 ± 3.61 (SD); 469 male, mean age 27.87 ± 3.65) out of the 1200 Subjects Data Release of the Human Connectome Project (HCP) [26] were analyzed. The current study utilized both minimally preprocessed and FIX ICA cleaned BOLD resting-state data [15] acquired over four 15-min multiband BOLD resting-state scans over 2 days. Only subjects completing all four resting-state scans are included in this analysis. The first 20 volumes of each run were excluded, yielding 1180 timepoints.

Region of Interest (ROI) Selection. Gray matter regions of interest consisted of 333 regions in the cerebral cortex [14], 14 subject-specific subcortical regions from FreeSurfer derived segmentation [12] (bilateral thalamus, caudate, putamen, amygdala, hippocampus, pallidum and nucleus accumbens) and 14 bilateral cerebellar representations of a 7-network parcellation [28]. This combined parcellation scheme incorporates gray matter ROIs totalling 361 regions.

Image Processing. A time series for each scan in each subject was extracted from FIX ICA cleaned and minimally preprocessed BOLD data. The minimally preprocessed BOLD data were also analyzed with head motion, white matter and CSF regression and these regressors plus global signal regression [25].

Functional Connectivity Calculation. Functional connectivity was calculated in both time and space domains (Supplement, Fig. 5). For the space domain, a 361×361 matrix was computed for each scan representing the Pearson correlation coefficient between the time series for each pair of gray matter regions. For the time domain, the matrix of 361×1180 time series was transposed and the correlation coefficient was analogously calculated between 1180×1180 pairs of timepoints. Given the large number of intercorrelated node pairs, full correlation was used because of the potential instability of partial correlation results.

Graph-Theoretic Metrics. We selected four graph-theoretic metrics for comparison of reproducibility of results across preprocessing strategies and between time and space connectivity measurements: modularity, chacteristic path length, global efficiency and clustering coefficient. These were computed using the Brain Connectivity Toolbox software for Matlab [23].

Reproducibility Metrics. As a metric of reproducibility, the intraclass correlation coefficient (ICC) was calculated using the ICC.m function for Matlab using '1-k' parameter across four scans for each of 1003 subjects for each measurement. This represents the expected ICC value that would be obtained for four scans per subject (1 h of resting state fMRI data). To interpret the reproducibility of an ICC score, we used the following guidelines – Poor - less than 0.4; Fair 0.4–0.59; Good 0.6–0.74; Excellent 0.74–1 – according to [6].

Persistent Homology Analysis. To apply persistent homology to brain networks, we map a given brain network to a point cloud in a metric space, where network nodes map to points and the measures of association between pairs of nodes map to distances between pairs of points [27]. In this paper, the association between two nodes u, v in the brain network is measured by their correlation coefficients $corr(u, v)$. The idea is to map this association to a distance measure such that higher correlations between nodes map to smaller distances. We use the mapping $d(u, v) = \sqrt{1 - corr(u, v)}$. Subsequently, a nested sequence of Vietoris-Rips complexes (a type of simplicial complex) is constructed in the metric space for persistent homology computation. Dimension 0 persistence barcodes were calculated with the R toolbox package TDA [11]. Specifically, connectivity matrices were converted into normalized distance matrices and used directly as input into function ripsDiag from the TDA library.

Cognitive and Personality Variables. Subject-level cognitive and personality scores used in this study contained scores from the 12 cognition domain measures included in the HCP battery of behavioral and individual difference measures - Cognition Domain [2] and 5 factor-level scores for personality from the NEO Five Factor Inventory (NEO FFI) [7]. We use corrected scoring of Agreeableness factor rather than the initial data supplied with the 1200 subjects release of the Human Connectome Project dataset [8]. Specific cognitive measures included: Episodic Memory (Picture Sequence Memory), Executive Function/Cognitive Flexibility (Dimensional Change Card Sort), Executive Function/Inhibition (Flanker Inhibitory Control and Attention Task), Fluid Intelligence (Penn Progressive Matrices), Language/Reading Decoding (Oral Reading Recognition), Language/Vocabulary Comprehension (Picture Vocabulary), Processing Speed (Pattern Comparison Processing Speed), Self-regulation/Impulsivity (Delay Discounting), Spatial Orientation (Variable Short Penn Line Orientation Test), Sustained Attention (Short Penn Continuous Performance Test), Verbal Episodic Memory (Penn Word Memory Test), Working Memory (List Sorting).

4 Results

Effects of Image Preprocessing on Reproducibility of Functional Connectivity. Functional connectivity was calculated by computing Pearson correlation coefficients for each pair of fMRI time series for 1003 subjects, each with four 15 min scans, using 4 preprocessing strategies for each pair of 361 gray matter nodes. For each preprocessing strategy, reproducibility was calculated by computing the intraclass correlation coefficient (ICC) for each pair of gray matter regions, obtained from 1003 subjects and 4 scans per subject (Supplement, Fig. 4). Statistical analysis using 1-way ANOVA demonstrated that each set of ICC values was significantly different for all four preprocessing strategies ($F < 1.0e^{-7}$), with markedly higher ICC values for Independent Component nuisance regression and lowest ICC values for preprocessing including global signal regression. For ICA-based nuisance regression, ICC was excellent (>0.7) for almost all connections but weaker for connections involving the subcortex [25].

Reproducibility was also calculated for four graph-theoretic metrics and time and space topological persistent homology measures. For graph-theoretic and topological metrics obtained by functional connectivity over space and time, reproducibility was highest for independent component-based nuisance regression, shown in Fig. 2. For persistent homology measures in the spatial domain (Fig. 2, top right), the difference was striking, with markedly improved reproducibility using ICA-based regression compared to more limited preprocessing strategies, and the weakest reproducibility obtained when including global signal regression. Taken together, these results provided strong evidence that for functional connectivity, whether computed over space or time, and including both graph-theoretic and topological measures computed from functional

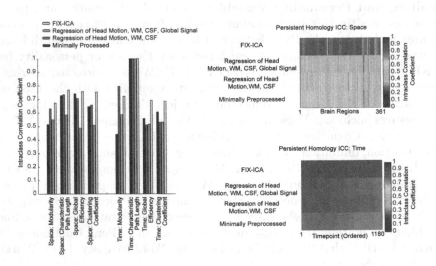

Fig. 2. Reproducibility of graph-theoretic and topological measures of functional connectivity by preprocessing strategy.

connectivity, the highest reproducibility was obtained using ICA-based nuisance regression. For all remaining analyses in this report, we used only FIX ICA cleaned data.

Differences in Results for Space and Time Connectivity. Reproducibility, as shown in Fig. 2, left, shows a comparable intraclass correlation for graph-theoretic measures obtained from connectivity with nodes representing spatial regions and with nodes representing timepoints. Reproducibility for global efficiency in time was higher than for other measures, which were all about 0.7.

Yet, when these graph-theoretic measures were compared across subjects to scores on 12 cognitive tests and 5 personality factors, only functional connectivity obtained from the time domain (using timepoints as nodes and correlation across spatial regions as edges) showed significant partial correlation to cognitive tests. Partial correlations were computed for each cognitive and personality metric separately, with age, sex and mean head motion used as subject-level covariates in each case, false discovery rate corrected. In particular, modularity in the time connectivity graphs was correlated with inhibitory components of executive function, and the characteristic path length and median clustering coefficient in the time connectivity graphs were correlated with vocabulary and processing speed (Supplement, Fig. 6).

Differences in Reproducibility and Behavioral Correlations for Persistent Homology. Persistence barcodes were calculated for connectivity graphs in the time (correlation across spatial ROIs) and space (correlation across timepoints) domains by calculating the "connectivity distance" at which each node merged with another cluster as described in Sect. 3. Each barcode consisted of a vector of 361 elements (space domain) or 1180 elements (time domain). Barcodes were reordered by persistence (length) for display in Fig. 3, left. Re-ordered barcodes were also used for analysis in the time domain, because timepoints are arbitrary and convey no consistent meaning across subjects or scans. In the space domain, barcodes were used without reordering, as each element of a barcode corresponds to a preserved brain region across subjects and scans.

The reproducibility of barcode results is shown in Fig. 3, center panels for space and time domains, yielding excellent (>0.7) ICC values for almost all elements. This is shown graphically on a template brain for the space domain as an inset in the figure, demonstrating that regions of lower ICC are exclusively in the medial orbitofrontal and medial anterior temporal regions, areas that are in close to brain/bone interfaces and are known to represent regions of high susceptibility artifact in the fMRI BOLD signal.

When comparing cognitive and personality metrics to persistence barcodes, also with partial correlation with age, sex and head motion as covariates, there were significant corrected correlations between persistent homology and fluid intelligence in both time and space domains, with less sensitivity to head motion when calculated in the time domain. For the spatial domain, fluid intelligence was correlated with barcode values in brain regions comprising association cortex of the frontal, parietal and temporal lobes, with weaker correlations in sensory and motor regions. Of 12 cognitive tests performed, 11 showed significant

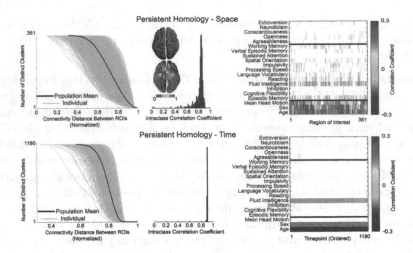

Fig. 3. Reproducibility and cognitive correlation of persistent homology barcode analyses.

partial correlation with persistent homology in the space domain after correction for multiple comparisons, and of five personality factors, four showed significant corrected partial correlation with persistent homology in the space domain (Supplement, Fig. 8 to Fig. 21, spatial distribution of correlation between persistent homology barcodes and specific cognitive measures including episodic memory, cognitive flexibility, agreeableness, openness, etc.)

5   Discussion

Topological data analysis of resting state functional connectivity using persistence barcodes identified individual differences in cognition and personality in the Human Connectome Project sample. Whether examining functional connectivity in the time or space domain, fluid intelligence was predicted by persistence barcode values, and distinct patterns of spatial regions were significantly correlated with a wide array of cognitive performance scores. Persistent homology showed excellent reproducibility across scans for the same individual for functional connectivity over both space and time in the brain. Graph-theoretic values also demonstrated good to excellent reproducibility, with functional connectivity in time and space domains correlated with distinct aspects of cognitive performance. For all tests, reproducibility was highest when using a robust independent component analysis nuisance regression strategy with a less connected graph than for other cleaning pipelines, suggesting that the spurious artifactual correlation between brain regions had been reduced. Reproducibility was lowest when using a strategy that includes global signal regression, possibly representing effects of contaminating results by incorporating information from other brain regions [20].

Not only were persistence barcodes significantly correlated with a surprising number of cognitive and personality features, but the spatial distributions of regions correlated with each behavior were also informative. Brain regions showing a correlation between barcode values and fluid intelligence were located in the association cortex, particularly in frontoparietal attentional regions, but not the sensory and motor cortex (Supplement, Fig. 7). These frontoparietal regions have been favored in the literature as substrates for general intelligence [16].

Similarly, correlations between persistent homology and working memory were observed in the ventral attention network and prefrontal cortex, the core neural substrates associated with pattern recognition, working memory and focused attention. A correlation between persistent homology and spatial attention specifically identified right hemispheric frontoparietal regions, consistent with well-known right dominance of lesions contributing to hemispatial neglect and lateralization of brain function to the right for spatial attention [21]. Both reading and language vocabulary scores were correlated with areas of the posterior temporal lobe associated with the Wernicke Area. Both episodic memory and verbal episodic memory scores were significantly correlated with areas of the posterior cingulate and medial temporal lobe, critical regions well known for their involvement in memory recall. Agreeableness was significantly correlated with persistent homology in the superior temporal sulcus, a core region of the social brain related to social empathy [22].

Although less intuitive than traditional functional connectivity between brain regions, our results suggest that functional connectivity between timepoints may offer new insights into aspects of cognition and neuropathology. Persistent homology, including potential higher dimensional topological features, may represent distinct aspects of brain function. These approaches may reflect dynamical aspects of connectivity such as the temporal duration and frequency of brain microstates or oscillations between metastable patterns of relative brain activity and provide new insights into brain network architecture or opportunities for the prediction of behavioral traits.

Acknowledgements. This work was supported by NIH grants R01EB022876, R01MH080826, and NSF grant IIS-1513616.

References

1. Allen, E., et al.: Tracking whole-brain connectivity dynamics in the resting state. Cereb. Cortex **24**(3), 663–676 (2009)
2. Barch, D., et al.: Function in the human connectome: task-fMRI and individual differences in behavior. NeuroImage **80**, 169–189 (2013)
3. Carlsson, G.: Topology and data. Bull. Am. Math. Soc. **46**, 255–308 (2009)
4. Carstens, C.J., Horadam, K.J.: Persistent homology of collaboration networks. Math. Prob. Eng. **2013** (2013)
5. Chen, J.E., Chang, C., Greicius, M.D., Glover, G.H.: Introducing co-activation pattern metrics to quantify spontaneous brain network dynamics. NeuroImage **111**, 476–488 (2015)

6. Cicchetti, D.: Guidelines, criteria, and rules of thumb for evaluating normed and standardized assessment instruments in psychology. Psychol. Assess. **6**(4), 284–290 (1994)
7. Costa, P.T., McCrae, R.R.: Revised NEO Personality Inventory (NEO-PI-R) and NEO Five-Factor Inventory (NEO-FFI) manual. In: Psychological Assessment Resources (1992)
8. Dubois, J., Galdi, P., Paul, L.K., Adolphs, R.: A distributed brain network predicts general intelligence from resting-state human neuroimaging data. bioRxiv (2018). https://doi.org/10.1101/257865
9. Edelsbrunner, H., Harer, J.: Computational Topology: An Introduction. American Mathematical Society, Providence (2010)
10. Edelsbrunner, H., Letscher, D., Zomorodian, A.J.: Topological persistence and simplification. Discrete Comput. Geom. **28**(4), 511–533 (2002)
11. Fasy, B.T., Kim, J., Lecci, F., Maria, C.: Introduction to the R package TDA. ArXiv:1411.1830 (2015)
12. Fischl, B., et al.: Whole brain segmentation: automated labeling of neuroanatomical structures in the human brain. Neuron **33**(3), 341–355 (2002)
13. Ghrist, R.: Barcodes: the persistent topology of data. Bull. Am. Math. Soc. **45**, 61–75 (2008)
14. Gordon, E., et al.: Generation and evaluation of a cortical area parcellation from resting-state correlations. Cereb. Cortex **16**(1), 288–303 (2016)
15. Griffanti, L., et al.: ICA-based artefact removal and accelerated fMRI acquisition for improved resting state network imaging. Neuroimage **95**, 232–247 (2014)
16. Jung, R., Haier, R.: The parieto-frontal integration theory (P-FIT) of intelligence: converging neuroimaging evidence. Behav. Brain Sci. **30**(2), 135–154 (2007)
17. Karahanoğlu, F., Van De Ville, D.: Transient brain activity disentangles fMRI resting-state dynamics in terms of spatially and temporally overlapping networks. Nat. Commun. **6**, 7751 (2015)
18. Keilholz, S., Caballero-Gaudes, C., Bandettini, P., Deco, G., Calhoun, V.: Time-resolved resting-state functional magnetic resonance imaging analysis: Current status, challenges, and new directions. Brain Connectivity **7**(8), 465–481 (2017)
19. Lee, H., Chung, M.K., Kang, H., Kim, B.N., Lee, D.S.: Discriminative persistent homology of brain networks. In: IEEE International Symposium on Biomedical Imaging: From Nano to Macro, pp. 841–844 (2011)
20. Murphy, K., Fox, M.D.: Towards a consensus regarding global signal regression for resting state functional connectivity MRI. NeuroImage **154**, 169–173 (2017)
21. Nielsen, J.A., Zielinski, B.A., Ferguson, M.A., Lainhart, J.E., Anderson, J.S.: An evaluation of the left-brain vs. right-brain hypothesis with resting state functional connectivity magnetic resonance imaging. PLoS One **8**(8), e71275 (2013)
22. Pelphrey, K., Morris, J., McCarthy, G.: Grasping the intentions of others: the perceived intentionality of an action influences activity in the superior temporal sulcus during social perception. J. Cogn. Neurosci. **16**(10), 1706–1716 (2004)
23. Rubinov, M., Sporns, O.: Complex network measures of brain connectivity: uses and interpretations. NeuroImage **52**(3), 1059–1069 (2010)
24. Saggar, M., et al.: Towards a new approach to reveal dynamical organization of the brain using topological data analysis. Nat. Commun. **9**(1), 1399 (2018)
25. Shah, L.M., Cramer, J.A., Ferguson, M.A., Birn, R.M., Anderson, J.S.: Reliability and reproducibility of individual differences in functional connectivity acquired during task and resting state. Brain Behav. **6**(5), e00456 (2016)
26. Van Essen, D., et al.: The WU-Minn human connectome project: an overview. Neuroimage **80**, 62–79 (2013)

27. Wong, E., et al.: Kernel partial least squares regression for relating functional brain network topology to clinical measures of behavior. In: International Symposium on Biomedical Imaging (2016)
28. Yeo, B., et al.: The organization of the human cerebral cortex estimated by intrinsic functional connectivity. J. Neurophysiol. **106**(3), 1125–1165 (2011)
29. Zomorodian, A., Carlsson, G.: Computing persistent homology. Discrete Comput. Geom. **33**(2), 249–274 (2005)

Riemannian Regression and Classification Models of Brain Networks Applied to Autism

Eleanor Wong[(⊠)], Jeffrey S. Anderson, Brandon A. Zielinski,
and P. Thomas Fletcher

University of Utah, Salt Lake City, UT 84112, USA
eleswong@sci.utah.edu

Abstract. Functional connectivity from resting-state functional MRI (rsfMRI) is typically represented as a symmetric positive definite (SPD) matrix. Analysis methods that exploit the Riemannian geometry of SPD matrices appropriately adhere to the positive definite constraint, unlike Euclidean methods. Recently proposed approaches for rsfMRI analysis have achieved high accuracy on public datasets, but are computationally intensive and difficult to interpret. In this paper, we show that we can get comparable results using connectivity matrices under the log-Euclidean and affine-invariant Riemannian metrics with relatively simple and interpretable models. On ABIDE Preprocessed dataset, our methods classify autism versus control subjects with 71.1% accuracy. We also show that Riemannian methods beat baseline in regressing connectome features to subject autism severity scores.

1   Introduction

Resting-state functional MRI (rsfMRI) has shown to be a promising imaging modality for diagnosing neurodevelopmental and neurodegenerative diseases, e.g., autism spectrum disorder (ASD) and Alzheimer's disease, and identifying associated biomarkers. However, analyses of imaging studies suffer from issues of low sample sizes, such that the conclusions are often not generalizable across datasets. The Autism Brain Imaging Data Exchange (ABIDE I) dataset is a joint effort from multiple international groups to aggregate a large dataset of imaging and phenotypic data for the purpose of identifying biomarkers of autism. To address heterogeneity in multisite data, the Preprocessed Connectome Project uses state of the art preprocessing that has shown good generalizability to the whole ABIDE I cohort [7]. This has fostered new methods for machine learning on covariance/correlation matrices of the preprocessed data.

Several recently proposed methods use deep neural networks (DNN) [2, 9, 11, 15, 17] to classify autism, achieving high accuracy. DNN learns a nonlinear mapping to semantically separate the data, but comes at the expense of high

© Springer Nature Switzerland AG 2018
G. Wu et al. (Eds.): CNI 2018, LNCS 11083, pp. 78–87, 2018.
https://doi.org/10.1007/978-3-030-00755-3_9

computation cost and difficult interpretability. These proposed methods do not take into account the SPD properties of correlation matrices.

Correlation matrices are symmetric semi-positive definite, and can be made symmetric positive definite (SPD) with a simple regularization step. The space of SPD matrices forms a Riemannian manifold. Using Euclidean operations on the manifold can be problematic, but many machine learning algorithms are only designed for the Euclidean space features. The two most commonly used Riemannian metrics proposed for the SPD manifold are the affine-invariant metric (AIM) and the log-Euclidean metric (LEM). The AIM is based on Lie group action on points on the SPD manifold, defined by a base point such that all other points are compared relative to. The LEM is equivalent to a special case of the AIM for which the base point is at identity, mapping the SPD manifold to the Euclidean space. These frameworks have been applied to brain network analyses in multiple studies. Varoquaux et al. [20] introduced a probabilistic model based on the AIM for comparing single subject correlation matrices from a group model to identify outlier stroke patients from a group of healthy controls. Ng et al. [16] used the AIM for transport on the SPD manifold to remove nonlinear commonalities between scans in longitudinal studies. Other works use the LEM to define kernels on the manifold for machine learning algorithms [8,23].

1.1 Contribution

Although works mentioned above have studied brain connectivity representations as SPD matrices on a Riemannian manifold, to the best of our knowledge, no one has demonstrated the performance of Riemannian methods on a ubiquitously used benchmark dataset such as ABIDE. Furthermore, regression between Riemannian representations of brain networks with neuropsychiatric features has not been explored. In our first contribution, we show that classification with a simple logistic regression using log mapped correlation matrices under the LEM achieves comparable results to other state-of-the-art deep neural network methods on the ABIDE dataset, with an accuracy of 70.0%. It uses a simple classification method (logistic regression) with little parameter tuning or engineering tricks. Due to the linearity of the classifier decision boundary, and the fact that log-Euclidean correlations retain the interpretatibility of the original correlations between pairs of regions, we can visualize the resulting classifier. Our second contribution is to show that the AIM can improve upon this accuracy, by proposing an optimization over the base point that yields a better performance at 71.1% accuracy.

2   Methods

The typical pipeline for rsfMRI analysis begins with the estimation of network as a connectome matrix using some measure of functional similarity between all pairs of regions of interest (ROIs) in the brain. To use the connectome for diagnosis of autism spectral disease, features are extracted from the correlation matrix as input into machine learning algorithms for classification. For many correlation-based measures, such as the most commonly used Pearson correlation, the matrices are symmetric semi-positive definite matrices. Thresholding the eigenvalues by some positive epsilon regularizes these correlation matrices to SPD.

We first review AIM and LEM, and then go over our preprocessing steps on the ABIDE dataset.

2.1   SPD Matrices

A $d \times d$ matrix M is symmetric positive definite if $z^T M z > 0$, $\forall z \neq 0 \in \mathbb{R}^d$. The space of all SPD matrices, denoted S_{++}^d, is not a vector space, but a Riemannian manifold. Using Euclidean operations on the manifold can be problematic, leading to the swelling effect, see e.g., [4]. Several metrics have been proposed for the SPD manifold [3,4,10,18]. Geodesic distance under the AIM [4,10], given by

$$\text{dist}\,(M_1, M_2) = \left\| M_1^{\frac{1}{2}} \log \left(M_2^{-\frac{1}{2}} M_1 M_2^{-\frac{1}{2}} \right) M_1^{\frac{1}{2}} \right\|_F,$$

addresses these issues. Under this Riemannian framework, two operations are introduced, the Riemannian exponential map and the Riemannian logarithmic map:

$$\text{Exp}_{M_1}\,(X) = M_1^{\frac{1}{2}} \exp \left(M_1^{-\frac{1}{2}} X M_1^{-\frac{1}{2}} \right) M_1^{\frac{1}{2}}$$

$$\text{Log}_{M_1}\,(M_2) = M_1^{\frac{1}{2}} \log \left(M_1^{-\frac{1}{2}} M_2 M_1^{-\frac{1}{2}} \right) M_1^{\frac{1}{2}},$$

where Exp and Log denote the Riemannian operations, and exp and log denote the matrix exponential and logarithm. $\text{Exp}_{M_1}\,(X)$ returns a point at time one along the geodesic starting at $M_1 \in S_{++}^d$ and with initial velocity vector X. $\text{Log}_{M_1}\,(M_2)$ is the inverse operation which yields that vector in the tangent space that Exp maps M_1 to M_2. For data analysis, consider M_1 as the base point that all data points are compared to. For example, M_1 can be set as the Fréchet mean, such as in [16].

Another proposed metric is the LEM [3], given by

$$\text{dist}\,(M_1, M_2) = \left\| \log\,(M_1) - \log\,(M_2) \right\|_F.$$

Notice that distances under the LEM are equivalent to those under the AIM when one of the two matrices, M_1 or M_2, is equal to the identity matrix. This becomes a way of mapping SPD matrices to the Euclidean tangent space at identity, i.e., $f : S_{++}^d \to \mathbb{R}^{d \times d}$

$$f_{\text{LEM}}\,(M) = \log\,(M). \tag{1}$$

After transforming data in S_{++}^d via the log map, we can apply Euclidean models, e.g., logistic regression. In the AIM case, the mapping of M_2 with respect to some basepoint M_1 is

$$f_{\text{AIM}}\,(M_2) = \log \left(M_1^{-\frac{1}{2}} M_2 M_1^{-\frac{1}{2}} \right). \tag{2}$$

We have the choice of either fixing the base point M_2 to the Fréchet mean and proceeding with Euclidean methods, or learning the base point simultaneously during optimization to select the best base point for the learning task. The reader may refer to [10] for the computation of the Fréchet mean for the SPD manifold under AIM.

For the optimization over the base point, we propose to use the backpropagation computation for matrix operations, i.e., matrix logarithm, described in [12,13]. As a concrete example, in the 2-class logistic regression case, the probability of a data point M_2 being in class Y is given by

$$P(Y = 1|M_2) = \frac{1}{1 + \exp\left(-\left(c + \beta\text{vec}\left(\log\left(M_1^{-1/2}M_2M_1^{-1/2}\right)\right)\right)\right)},$$

where M_1 is a base point to optimize over and vec $(\cdot)$ is the vectorization of a matrix. The energy function is the standard cross-entropy for logistic regression. Because of chain rule, minimizing the cross-entropy with respect to M_1 involves computing the matrix logarithm backgradient Z, using a neural network-like setup such that the matrix logarithm is a "layer" upon its inputs (refer to [12] for details). Afterwards, the gradient with respect to $C = M_1^{-1/2}$ is $\nabla_C = M_2CZ + ZCM_2$. We can update the base point in a couple of ways: (1) by standard (additive) gradient descent and then regularizing the resulting M_1 to have all positive eigenvalues, or (2) by taking the Exp map, $C_{new} = \text{Exp}_{C_{old}}(\nabla_{C_{old}})$.

2.2   ABIDE

The ABIDE I dataset is a collection of rsfMRI and phenotypic data for typically developing controls and ASD subjects acquired at 20 different sites. The Preprocessed Connectome Project [7] has preprocessed ABIDE data using state of the art pipelines to promote shareability and fair comparison of results. We obtain the fMRI data from the Project, preprocessed with the CPAC pipeline and parcellated according to the Harvard-Oxford atlas, and select the 871 subjects (468 controls, 403 ASD) to be consistent with [1,17]. The resulting time series at each of the $d = 111$ regions are normalized to mean = 0 and standard deviation = 1.

3   Results

3.1   Classification

We first compare between raw and Fisher-transformed Pearson's correlation matrices, as well as eigenvalue-regularized and log-Euclidean transformed matrices as input for each subject into logistic regression for classification. We eigendecompose raw correlation matrices and lower-bound small eigenvalues to 0.5, and re-compose them into regularized correlation matrices to ensure that the matrices are SPD. Log-Euclidean matrices are obtained by taking the matrix logarithm of the regularized correlation matrices. All matrices are then reduced to upper triangles and vectorized into feature vectors. Matrix features involving log-Euclidean transform are of 6205 dimensions because diagonal entries are included in the upper triangle, whereas all other features are of 6105 dimensions.

We use the Scikit-Learn implementation of logistic regression with L_2 penalty as classifier, and evaluate the classification performance through a nested ten-fold cross-validation scheme (folds selected at random). At each fold, 10% of the data is set aside for testing, and the other 90% is ten-fold cross-validated to get the best parameter for L_2 penalty. The range of parameters we cross-validate over are [0.01, 0.05, 0.075, 0.1, 0.2, 0.5, 0.75, 1.0, 3.0, 5.0].

We then also compare the affine-invariant transformed matrices with an optimization for the base point in TensorFlow using the same ten-fold cross-validation scheme. At the first layer, square correlation matrices are affine-invariant transformed with variable M_2, then linearized to a 6205 dimensional vector and fed into a sigmoid function for classification. The cost function to optimize over is the sum of the logistic regression cross-entropy plus L_2 penalty with parameter λ. In TensorFlow, the range of parameters we cross-validate over are [5, 10, 15, 20, 100, 200]. The matrix backpropagation is modified to the method described in the previous section. The optimization is run until convergence within 50 iterations.

Table 1 shows the results. Our baseline of using just vectorized correlation matrix features has an accuracy score of 65.7%, comparable to baseline scores reported in [1,17]. A t-test shows that both the log-Euclidean and the affine-invariant transformed features have a statistical significant improvement in performance over the raw correlation baseline ($p = 0.02$ and $p = 0.002$, respectively). The regularized correlation matrix shows similar accuracy to the raw correlation features, indicating that the increase in performance is solely due to the Riemannian mappings. Figure 2 describes the range of classification accuracy from ten-fold cross-validation for the baseline compared to log-Euclidean and affine-invariant mapped features. Using the model learned from the log-Euclidean features, we visualize the highest weights in the classification thresholded at $|w| > 0.25$ in Fig. 3. Red connections indicate positive weights that push classification toward the ASD group (label = 1) and blue connections are negative weights toward the control group.

Figure 1 is a diagram of the learned weights on the ROIs grouped by subnetworks from [19]–visual, default mode, sensorimotor, auditory, executive control, and frontoparietal networks. The colormap runs from negative values in blue (driving classification toward control) to positive values in red (toward autism). For visualization, very small weights have been filtered. There is evidence of patterns within and between subnetworks. The weights within a subnetwork are simplified to their means within a block. Our results show that control subjects tend to have higher intranetwork connectivity especially within the sensorimotor, executive control, and default mode networks, whereas subjects with ASD have stronger internetwork connectivity, e.g., between the default mode and the

Table 1. Accuracy performance of Riemannian and various state of the art classification methods

Method	Validation	Accuracy (Stdev)	Sensitivity	Specificity
Abraham et al. [1]	CV10	0.668	-	-
Dvornek et al. [9]	CV10	0.685 (0.06)	-	-
Parisot et al. [17]	CV10	0.695	-	-
Heinsfeld et al. [11]	CV10	0.70	0.74	0.63
Raw correlation	CV10	0.657 (0.06)	0.728	0.573
Fisher correlation	CV10	0.672 (0.05)	0.737	0.594
Regularized correlation	CV10	0.660 (0.06)	0.741	0.565
Log-Euclidean	CV10	0.700 (0.05)	0.809	0.575
Affine-Invariant	CV10	**0.711 (0.05)**	0.838	0.585

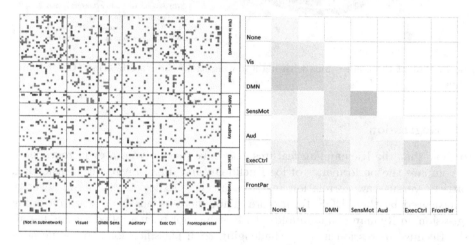

Fig. 1. Classification weights grouped by subnetwork (Color figure online)

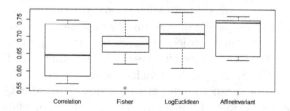

Fig. 2. Box plots of the classification accuracy over ten-fold cross-validation.

sensorimotor networks. This is in agreement with existing literature that the default mode network is not well segregated from other subnetworks for the ASD population [5,21,22].

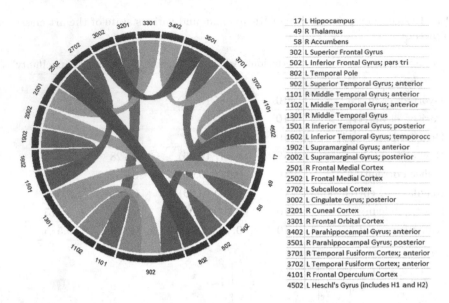

17	L Hippocampus
49	R Thalamus
58	R Accumbens
302	L Superior Frontal Gyrus
502	L Inferior Frontal Gyrus; pars tri
802	L Temporal Pole
902	L Superior Temporal Gyrus; anterior
1101	R Middle Temporal Gyrus; anterior
1102	L Middle Temporal Gyrus; anterior
1301	R Middle Temporal Gyrus
1501	R Inferior Temporal Gyrus; posterior
1602	L Inferior Temporal Gyrus; temporocc
1902	L Supramarginal Gyrus; anterior
2002	L Supramarginal Gyrus; posterior
2501	R Frontal Medial Cortex
2502	L Frontal Medial Cortex
2702	L Subcallosal Cortex
3002	L Cingulate Gyrus; posterior
3201	R Cuneal Cortex
3301	R Frontal Orbital Cortex
3402	L Parahippocampal Gyrus; anterior
3501	R Parahippocampal Gyrus; posterior
3701	R Temporal Fusiform Cortex; anterior
3702	L Temporal Fusiform Cortex; anterior
4101	R Frontal Operculum Cortex
4502	L Heschl's Gyrus (includes H1 and H2)

Fig. 3. Plot of the connections with highest weights in the classification (Color figure online)

3.2   Regression

To show that the Riemannian features also have predictive power in regression, we compare the performance of log-Euclidean and affine-invariant transformed matrices versus raw correlation matrices in the prediction of autism severity as measured by the ADOS Total score. Though there has been work on doing regression on Riemannian manifolds [6,14], it has not been applied for ASD analysis. Because regression is more challenging than classification, and some sites lack ADOS scores for control subjects, we limit our analysis to the largest site with a roughly even split of ASD and control subjects that have ADOS Total score. The Utah site has 62 subjects with scores ranging from 0 to 21. Scores below 10 are considered typically developing. We use partial least squares regression (PLS), with the features projected down to one component and regressed to ADOS. It is not trivial to adapt the base point optimization for the NIPALS algorithm in solving PLS. Instead, here we fix the base point to the Fréchet mean. Table 2 shows the root mean squared error (RMSE), R^2 and Q^2 coefficient of determination values between Riemannian and baseline correlation matrix features. The R^2 is computed over the whole data subset, and the Q^2 value is calculated through a leave-one-out cross-validation (LOOCV) scheme. The plot of true versus predicted ADOS using the Fréchet mean base point is shown in Fig. 4. To show the statistical significance of improvement, we do a permutation test. We sum up the absolute value of the residuals of the LOOCV predictions and take the difference of the proposed method from the baseline correlation as the test statistic. Then we do 10000 permutations swapping the predictions between the two classes and sum up the number of times that the differences are greater than our nonpermuted test statistic value. Both log-Euclidean and affine-invariant metrics significantly improve over the raw and Fisher correlation baselines (also similarly significant by t-test on the RMSE).

Table 2. Comparison of Riemannian features against baselines in PLS regression

	RMSE	R^2	Q^2	Raw Corr Improve	Fish Corr Improve
Raw correlation	6.17	0.631	−0.05	-	-
Fisher correlation	6.18	0.624	−0.062	-	-
Log-Euclidean	5.42	0.816	0.182	$p = 0.0112$	$p = 0.0127$
Affine-Invariant	**5.36**	**0.837**	**0.202**	$p = 0.0064$	$p = 0.0069$

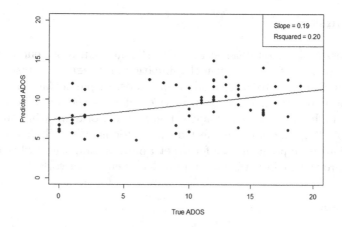

Fig. 4. Predicted vs. true ADOS scores for regression under AIM

The regression weights in Fig. 5 show similar patterns to the classification results, though not the same. This is expected because the regression data is only a single-site subset. The classification and the regression weights share a correlation of 0.31, reasonably consistent for such high-dimensional data. Summarizing weights into means of each block, we can see the pattern that the intra-connectivity in the default mode and sensorimotor networks drives the regression toward low ADOS scores (control) and interconnectivity between the two networks pushes regression toward high ADOS scores.

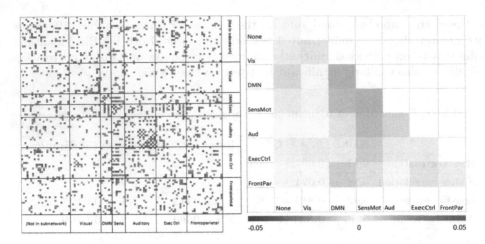

Fig. 5. Regression weights grouped by subnetwork

4   Conclusion

In this paper, we have established that the Riemannian representation of SPD matrices is beneficial for the autism classification and regression tasks and comparable in performance to other modern methods. In particular, the results are interpretable under the log-Euclidean metric, whereas the affine-invariant metric leads to high learning performance. For future work, we will compare how the choice of ROI may have an effect on predictions. We will also develop the affine-invariant base point update for other analyses, and study whether it may yield an improvement in performance in a deep neural network.

References

1. Abraham, A., et al.: Deriving reproducible biomarkers from multi-site resting-state data: an autism-based example. NeuroImage **147**, 736–745 (2017)
2. Anirudh, R., Thiagarajan, J.J.: Bootstrapping graph convolutional neural networks for autism spectrum disorder classification. arXiv preprint arXiv:1704.07487 (2017)
3. Arsigny, V., Fillard, P., Pennec, X., Ayache, N.: Log-Euclidean metrics for fast and simple calculus on diffusion tensors. Magn. Reson. Med. **56**(2), 411–421 (2006)
4. Arsigny, V., Fillard, P., Pennec, X., Ayache, N.: Geometric means in a novel vector space structure on symmetric positive-definite matrices. SIAM J. Matrix Anal. Appl. **29**(1), 328–347 (2007)
5. Assaf, M., Jagannathan, K., Calhoun, V.D., Miller, L., Stevens, M.C., et al.: Abnormal functional connectivity of default mode subnetworks in autism spectrum disorder patients. NeuroImage **53**(1), 247–256 (2010)
6. Cornea, E., Zhu, H., Kim, P., Ibrahim, J.G., Alzheimers Disease Neuroimaging Initiative: Regression models on Riemannian symmetric spaces. Stat. Methodol. **79**(2), 463–482 (2017)

7. Craddock, C., et al.: The neuro bureau preprocessing initiative: open sharing of preprocessed neuroimaging data and derivatives
8. Dodero, L., Minh, H.Q., San Biagio, M., Murino, V., Sona, D.: Kernel-based classification for brain connectivity graphs on the Riemannian manifold of positive definite matrices. In: Biomedical Imaging (ISBI), pp. 42–45. IEEE (2015)
9. Dvornek, N.C., Ventola, P., Pelphrey, K.A., Duncan, J.S.: Identifying autism from resting-state fMRI using long short-term memory networks. In: Wang, Q., Shi, Y., Suk, H.-I., Suzuki, K. (eds.) MLMI 2017. LNCS, vol. 10541, pp. 362–370. Springer, Cham (2017). https://doi.org/10.1007/978-3-319-67389-9_42
10. Fletcher, P.T., Lu, C., Pizer, S.M., Joshi, S.: Principal geodesic analysis for the study of nonlinear statistics of shape. TMI **23**(8), 995–1005 (2004)
11. Sólon Heinsfeld, A., Franco, A.R., Craddock, R.C., Buchweitz, A., Meneguzzi, F.: Identification of autism spectrum disorder using deep learning and the abide dataset. NeuroImage Clin. **17**, 16–23 (2018)
12. Huang, Z., Van Gool, L.J.: A Riemannian network for SPD matrix learning (2017)
13. Ionescu, C., Vantzos, O., Sminchisescu, C.: Matrix backpropagation for deep networks with structured layers. In: ICCV, pp. 2965–2973 (2015)
14. Kim, H.J., et al.: Multivariate general linear models (MGLM) on Riemannian manifolds with applications to statistical analysis of diffusion weighted images. In: CVPR, pp. 2705–2712 (2014)
15. Meszlényi, R.J., Buza, K., Vidnyánszky, Z.: Resting state fMRI functional connectivity-based classification using a convolutional neural network architecture. Front. Neuroinformatics **11**, 61 (2017)
16. Ng, B., Dressler, M., Varoquaux, G., Poline, J.B., Greicius, M., Thirion, B.: Transport on Riemannian manifold for functional connectivity-based classification. In: Golland, P., Hata, N., Barillot, C., Hornegger, J., Howe, R. (eds.) MICCAI 2014. LNCS, vol. 8674, pp. 405–412. Springer, Cham (2014). https://doi.org/10.1007/978-3-319-10470-6_51
17. Parisot, S., et al.: Spectral graph convolutions for population-based disease prediction. In: Descoteaux, M., Maier-Hein, L., Franz, A., Jannin, P., Collins, D.L., Duchesne, S. (eds.) MICCAI 2017. LNCS, vol. 10435, pp. 177–185. Springer, Cham (2017). https://doi.org/10.1007/978-3-319-66179-7_21
18. Pennec, X., Arsigny, V., Fillard, P., Ayache, N.: Fast and simple computations on tensors with log-Euclidean metrics (2005)
19. Smith, S.M., Fox, P.T., Miller, K.L., et al.: Correspondence of the brain's functional architecture during activation and rest. PNAS **106**(31), 13040–13045 (2009)
20. Varoquaux, G., Baronnet, F., Kleinschmidt, A., Fillard, P., Thirion, B.: Detection of brain functional-connectivity difference in post-stroke patients using group-level covariance modeling. In: Jiang, T., Navab, N., Pluim, J.P.W., Viergever, M.A. (eds.) MICCAI 2010. LNCS, vol. 6361, pp. 200–208. Springer, Heidelberg (2010). https://doi.org/10.1007/978-3-642-15705-9_25
21. Weng, S., et al.: Alterations of resting state functional connectivity in the default network in adolescents with autism spectrum disorders. Brain Res. **1313**, 202–214 (2010)
22. Yerys, B.E., Gordon, E.M., Abrams, D.N., Satterthwaite, T.D., et al.: Default mode network segregation and social deficits in autism spectrum disorder: evidence from non-medicated children. NeuroImage Clin. **9**, 223–232 (2015)
23. Young, J., Lei, D., Mechelli, A.: Discriminative log-Euclidean kernels for learning on brain networks. In: Wu, G., Laurienti, P., Bonilha, L., Munsell, B.C. (eds.) CNI 2017. LNCS, vol. 10511, pp. 25–34. Springer, Cham (2017). https://doi.org/10.1007/978-3-319-67159-8_4

Defining Patient Specific Functional Parcellations in Lesional Cohorts via Markov Random Fields

Naresh Nandakumar[1]($\boxtimes$), Niharika S. D'Souza[1], Jeff Craley[1], Komal Manzoor[2], Jay J. Pillai[2], Sachin K. Gujar[2], Haris I. Sair[2], and Archana Venkataraman[1]

[1] Department of Electrical and Computer Engineering,
Johns Hopkins University, Baltimore, USA
nnandak1@jhu.edu
[2] Department of Neuroradiology, Johns Hopkins School of Medicine,
Baltimore, USA

Abstract. We propose a hierarchical Bayesian model that refines a population-based atlas using resting-state fMRI (rs-fMRI) coherence. Our method starts from an initial parcellation and then iteratively reassigns the voxel memberships at the subject level. Our algorithm uses a maximum *a posteriori* inference strategy based on the neighboring voxel assignments and the Pearson correlation coefficients between the voxel time series and the parcel reference signals. Our method is generalizable to different initial atlases, ensures spatial and temporal contiguity in the final network organization, and can handle subjects with brain lesions, whose rs-fMRI data varies tremendously from that of a healthy cohort. We validate our method by comparing the intra-network cohesion and the motor network identification against two baselines: a standard functional parcellation with no reassignment and a recently published method with a purely data-driven reassignment procedure. Our method outperforms the original functional parcellation in intra-network cohesion and both methods in motor network identification.

Keywords: Rs-fMRI · Markov Random Field
Patient-specific networks

1 Introduction

Resting-state fMRI (rs-fMRI) captures the intrinsic communication patterns in the brain. Neural activity at rest is organized into Resting State Networks (RSNs), which are determined by both spatial coherence and temporal synchrony at the voxel level [1]. Non-invasive methods for identifying RSNs are of particular interest when planning neurosurgery, where the goal is to localize (and subsequently avoid) cruicial motor and language functionality, also known as the eloquent cortex. Accurately identifying these key functional networks will determine safe resection margins and may also help us to predict the functional

© Springer Nature Switzerland AG 2018
G. Wu et al. (Eds.): CNI 2018, LNCS 11083, pp. 88–98, 2018.
https://doi.org/10.1007/978-3-030-00755-3_10

outcomes of surgery [2]. Task-fMRI, where the subject is performing an explict language or motor paradigm, is currently the most popular technique for mapping eloquent brain areas. However, recent interest has moved towards rs-fMRI to overcome both the unreliability of certain task activations and the inability of critically ill patients to complete a cognitively demanding protocol [2].

While a standard functional parcellation is often sufficient to identify RSNs in healthy individuals, a subject specific approach is necessary for brain lesion patients due to a compensatory mechanism known as neural plasticity. For example, it has been shown that motor and language functionality in patients with low-grade gliomas can migrate to other parts of the brain [3]. Therefore, data from glioma patients may not conform to a standard functional atlas.

Previous work has focused on deriving subject-specific functional atlases from the original rs-fMRI data. These techniques include, but are not limited to, spatially constrained hierarchical parcellations [4] and data-driven clustering techniques [5]. In contrast, we approach this problem as one of *atlas refinement* rather than atlas construction. This strategy allows us to work with *any* previously validated anatomical or functional parcellation. Liu et. al have introduced a method to obtain subject-specific RSNs by iteratively reassigning the voxel memberships based on the Pearson correlation coefficients between the voxel time series and a collection of network reference signals. This method was originally derived to determine functional network differences in healthy subjects and showed promise for clinical populations [6]. These reference signals are calculated as a weighted average of the previous iteration's reference signal and the current average time series for the RSN. Their method does not explicitly consider spatial contiguity in RSNs, and requires the user to specify various parameters.

We develop a Bayesian model that uses both spatial and temporal information to iteratively refine an initial functional parcellation on a patient-specific basis. Our model uses a Markov Random Field (MRF) prior to encourage spatial contiguity within the functional parcels. We employ a maximum *a posteriori* (MAP) inference strategy for voxel-wise network assignment until a predefined convergence criteria is met. Our method builds on prior work in Bayesian network modeling [7] and MRF priors for rs-fMRI data [8]. We validate our method on rs-fMRI data from 67 glioma patients. Our initial atlas is the Yeo 17 network functional parcellation [9], which is one of the most widely cited functional atlases in the literature. We compare the performance of our method with the original parcellation (no reassignment) and with reassignment according to Liu, whose paper references the same parcellation. Our validation metrics include the intra-network cohesion amongst the final RSNs and motor network identification via task based fMRI concordance using three distinct task paradigms.

2   A Bayesian Model for Voxel Reassignment

Our model infers an underlying (i.e. latent) network architecture that integrates both spatial contiguity and temporal synchrony across the brain. At each voxel, we leverage the time series data, the neighborhood network membership, and a

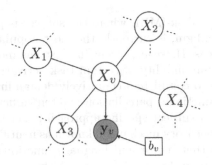

Fig. 1. A graphical model of our framework where shaded nodes represent observed random variables.

binary tumor label, indicating if the voxel lies within the lesion or not. Let X_v be the network assignment for voxel v. In our framework, X_v gets assigned to one of $K+1$ values, where K is the number of networks (or region parcels) defined in the initial atlas. An assignment of $X_v = 0$ indicates no network membership for voxel v if it belongs to the glioma. Mathematically, let $\boldsymbol{X}_{-v}$ be the current network assignments of all other voxels in the brain. Likewise, $\boldsymbol{y}_v$ is the time series data at voxel v, $\boldsymbol{\mu}_k$ is the current reference signal for network $k \in \{1 \dots K\}$, and $b_v \in \{0, 1\}$ is the binary tumor label such that $b_v = 0$ implies that voxel v is tumorous. Our setup is illustrated in Fig. 1. As seen, the assignment for X_v depends on its immediate spatial neighbors. The relationship between X_v and $\boldsymbol{X}_{-v}$ is captured by an MRF prior while the relationship between X_v and $\boldsymbol{y}_v, b_v$ is captured by the data likelihood. For visualization purposes the 2D representation in Fig. 1 shows four neighbors per pixel. However, we have implemented a 3D model, which has six neighbors per voxel.

We encourage spatial contiguity in our latent network assignments by stipulating that voxel v will be more likely to assume the state of its neighboring voxels. We model the MRF prior after the Potts model [10]:

$$P(X_v = k | \boldsymbol{X}_{-v}) = \frac{1}{Z_x} \Psi(X_v, \boldsymbol{X}_{-v}, k) \propto \left\{ 1 + \exp\left[-\left(\beta + \sum_{i \in ne(v)} \mathbb{1}_{X_i = k}\right) \right] \right\}^{-1} \tag{1}$$

where β controls the influence of the neighbor voxel network memberships on voxel v. Here, $ne(v)$ denotes the neighbors of voxel v, and the sum $\sum_{i \in ne(v)} \mathbb{1}_{X_i = k}$ captures how often these neighbors are assigned to network k. Notice that this sum will be zero for network k when $X_i \neq k$ for all $i \in ne(v)$.

The likelihood $P(\boldsymbol{y}_v | X_v = k; \boldsymbol{\mu}_k, b_v)$ is modeled after a rescaled version of the Pearson correlation coefficient between the reference $\boldsymbol{\mu}_k$ and data $\boldsymbol{y}_v$:

$$\rho = \frac{\text{cov}(\boldsymbol{y}_v, \boldsymbol{\mu}_k)}{\sigma_{\boldsymbol{y}_v} \sigma_{\boldsymbol{\mu}_k}} \tag{2}$$

where ρ is subsequently shifted and scaled to be between $[0, 1]$ to allow for a normalizable density. The final likelihood model is given by

$$P(\boldsymbol{y}_v | X_v = k; \boldsymbol{\mu}_k, b_v) = \frac{1}{Z_y} \Psi(\boldsymbol{y}_v, \boldsymbol{\mu}_k, b_v) = \frac{1}{Z_y} \left\{ \frac{(\rho_{\boldsymbol{y}_v, \boldsymbol{\mu}_k} + 1)}{2} \times b_v \right\}. \quad (3)$$

The rescaled Pearson correlation coefficient goes to one for strong positive correlations and zero for strong negative correlations. The label b_v sets the likelihood to zero for tumorous voxels, which corresponds to no network membership.

2.1 Approximate Posterior Inference

The observed rs-fMRI time series $\boldsymbol{y}_v$ are conditionally independent given $\{X_v\}$. Based on the model factorization, our posterior distribution can be written as

$$P(X_v = k | \boldsymbol{X}_{-v}, \boldsymbol{y}_v; \theta) = \frac{1}{Z} \Psi(X_v, \boldsymbol{X}_{-v}, k) \Psi(\boldsymbol{y}_v, \boldsymbol{\mu}_k, b_v) \quad (4)$$

where $\Psi(X_v, \boldsymbol{X}_{-v}, k)$ models the prior, $\Psi(\boldsymbol{y}_v, \boldsymbol{\mu}_k, b_v)$ models the likelihood under the belief that $X_v = k$, and Z is a normalization constant that combines both Z_x and Z_y. We have derived an update procedure based on maximizing the following log-posterior over all possible network assignments:

$$X_v^* = \underset{k}{\operatorname{argmax}} \left\{ -\log Z + \log \Psi(X_v, \boldsymbol{X}_{-v}, k) + \log \Psi(\boldsymbol{y}_v, \boldsymbol{\mu}_k, b_v) \right\}. \quad (5)$$

2.2 Implementation Details

We have derived an algorithm based on max product message passing to ensure atlas stability [11]. Our algorithm iterates between two main steps: updating the network assignments $\{\boldsymbol{X}_v\}$ and updating the reference signals $\{\boldsymbol{\mu}_k\}$. Let $\boldsymbol{Y} \in \mathbb{R}^{x \times y \times z \times T}$ be the aggregated rs-fMRI data across all (x, y, z) spatial coordinates and let $\boldsymbol{X}^{(t)} \in \mathbb{R}^{x \times y \times z}$ be the assignment information at iteration t. Let $\boldsymbol{B} \in \mathbb{R}^{x \times y \times z}$ be the binary tumor matrix. The values stored in $\boldsymbol{B}$ are 0 at tumorous voxels and 1 elsewhere. We initialize our algorithm with the Yeo atlas and then the Hadamard product $\boldsymbol{X}^{(0)} = \boldsymbol{X} \odot \boldsymbol{B}$, which defaults all tumorous voxel assignments to 0 due to unreliable signal at these locations. We then parcellate $\boldsymbol{Y}$ by the assignments in $\boldsymbol{X}^{(0)}$ and calculate the initial reference signals $\boldsymbol{\mu}_k^{(0)}$ for $k \in \{1 \ldots K\}$ with

$$\boldsymbol{\mu}_k^{(t)} = \frac{\sum_{v=1}^{V} \boldsymbol{Y}_v \cdot \mathbb{1}(\boldsymbol{X}_v^{(t)} = k)}{\sum_{v=1}^{V} \mathbb{1}(\boldsymbol{X}_v^{(t)} = k)}. \quad (6)$$

At each main iteration t, we determine the voxel assignments I times according to our MAP rule in Eq. (5) initializing with $\boldsymbol{X}^{(t-1)}$. Using the assignments in $i - 1$, we employ a flooding schedule to simultaneously determine the network values $\widehat{\boldsymbol{X}}^{(i)}$ at iteration i. The updated assignments $\boldsymbol{X}^{(t)}$ is given by majority

---

Algorithm 1. Max product message passing procedure for atlas refinement. Here, c is a prespecified threshold on the network consistency between iterations.

1: **procedure** MRFREFINEMENT$(\boldsymbol{X}, \boldsymbol{B}, \boldsymbol{Y})$
2:　　$\boldsymbol{X}^{(0)} \leftarrow \boldsymbol{X} \odot \boldsymbol{B}$
3:　　$\{\boldsymbol{\mu}_1^{(0)} \cdots \boldsymbol{\mu}_K^{(0)}\} \leftarrow \boldsymbol{Y}, \boldsymbol{X}^{(0)}$　　　　　　　　　　　　　　　$\triangleright$ Eq. (6)
4:　　$t \leftarrow 1$
5:　　$M \leftarrow 0$　　　　　　　　　　　　　　　　$\triangleright$ Membership retention $M \in [0,1]$
6:　　**while** $M < c$ **do**　　　　　　　　　　$\triangleright$ Convergence threshold $c \in [0,1]$
7:　　　　$\widehat{\boldsymbol{X}}^{(1)} \leftarrow \boldsymbol{X}^{(t-1)}$
8:　　　　**for** $i = 2 : I$ **do**
9:　　　　　　**for** $v \in V$ **do**
10:　　　　　　　$\widehat{\boldsymbol{X}}_v^{(i)} \leftarrow \text{argmax}_k \left\{ \log \Psi(X_v, \widehat{\boldsymbol{X}}_{-v}^{(i-1)}, k) + \log \Psi(\boldsymbol{y}_v, \boldsymbol{\mu}_k^{(t-1)}, b_v) \right\}$
11:　　　　$\boldsymbol{X}^{(t)} \leftarrow mode(\{\widehat{\boldsymbol{X}}^{(i)}\}_{i=1}^I)$
12:　　　　$\{\boldsymbol{\mu}_1^{(t)} \cdots \boldsymbol{\mu}_K^{(t)}\} \leftarrow \boldsymbol{Y}, \boldsymbol{X}^{(t)}$　　　　　　　　　　　　　　　$\triangleright$ Eq. (6)
13:　　　　$M \leftarrow \frac{\sum_v \mathbb{1}(X_v^{(t-1)} = X_v^{(t)}) \cdot B_v}{\sum_v B_v}$　　　$\triangleright$ Fraction of retained voxel memberships
14:　　　　$t \leftarrow t + 1$
15:　　**return** $\boldsymbol{X}^t$

---

vote over the determined network values $\{\widehat{\boldsymbol{X}}^{(i)}\}_{i=1}^I$. Given the new assignments, we update the network signals by Eq. (6) for $k \in \{1 \ldots K\}$ and check if the convergence criteria has been met by calculating the fraction of non-zero assigned voxels retaining the same network membership between iterations $t - 1$ and t. If the membership consistency between iterations is less than a specified stopping criteria, we repeat the procedure. Each voxel of interest has six neighbors as determined by adjacency in each of the three coordinate directions. Algorithm 1 presents our pseudo-code where the subject-specifc inputs are $\boldsymbol{B}$ and $\boldsymbol{Y}$.

2.3　Baseline Comparisons

We compare our Bayesian approach with the original parcellation and with the voxel reassignment method described by Liu [6]. We use the Yeo 17 network atlas due to its strong reproducibility and the large sample size used for construction. We confine our experiments to the more conservative cortical ribbon version of the Yeo atlas to get a more detailed parcellation. The method of Liu initializes the reference signals to the average time series defined by the original parcellation. From here, Liu reassigns voxel v by considering the maximum correlation between its time series and all K reference signals. A confidence value for each voxel is also computed as the ratio of the maximum correlation over the second highest correlation. The reference signal updates are only taken from voxels that have confidence values which exceed a predetermined threshold. They are computed as weighted combinations of the previous iteration's reference signals with the updated reference signals. The corresponding weights are nonlinear functions of the signal-to-noise ratio, the inter-subject variability, and the iteration number. We applied the Liu baseline with the parameter suggestions provided in [6], which were optimized for the 17 network Yeo atlas. We initialize each method, original, Liu, and proposed as described in Sect. 2.2.

3 Experimental Results

Dataset and Preprocessing: Our dataset includes task and rs-fMRI for 67 glioma patients who underwent preoperative mapping as part of their clinical workup. The fMRI were acquired using a 3.0 T Siemens Trio Tim (TR = 2000 ms, TE = 30 ms, flip angle = 90°, field of view = 24 cm, acquisition matrix = 64 × 64 × 33, slice thickness = 4 mm, and slice gap of 1 mm). We manually segmented each patient's tumor using MIPAV. Figure 2 illustrates tumor segmentations for three different patients to motivate the heterogeneity of our cohort. The fMRI was processed using SPM8. Both rs-fMRI and task-fMRI underwent slice timing correction, motion correction and registration to the MNI-152 template. The rs-fMRI was scrubbed using the ArtRepair toolbox in SPM, linearly detrended, underwent nuisance regression utilizing CompCor [12], bandpass filtered from 0.01 to 0.1 Hz, and spatially smoothed with a 6 mm FWHM Gaussian kernel.

General Linear Model (GLM) in SPM8 was used to derive activation maps for the three motor tasks [2]. Our dataset includes three different motor paradigms that were designed to target distinct parts of the motor homonculus [13]: finger tapping, tongue moving, and foot tapping. Since the task-fMRI data was acquired for clinical purposes, only 42 patients performed the finger task, 35 patients performed the tongue task, and 20 patients performed the foot task.

The population-based atlas contains 17 distinct functional networks confined to the cortical ribbon [9]. For both methods, a network retention convergence criteria of 0.98 was used. We chose $\beta = -0.5$ and $I = 100$ iterations for our model and a confidence value of 1.5 for the Liu baseline. Different combinations of reference signal calculation between updates for both our method and the Liu baseline were explored; we have reported the optimal results in each case.

3.1 Evaluating Resting State Network Cohesion

Our intra-network cohesion metric quantifies the temporal synchrony between voxels that belong to the same network [1]. Let V_k be the voxels assigned to

Fig. 2. The tumor segmentations (yellow) for three different patients are shown. (Color figure online)

network k, we define the Network Cohesion (NC) as the average correlation between voxels assigned to network k with the network signal μ_k.

$$NC_k = \frac{\sum_{j \in V_k} \rho_{y_j, \mu_k}}{|V_k|} \tag{7}$$

Figure 3 illustrates the difference in NC between our proposed method and both the original parcellation (left) and the Liu baseline (right). A value greater than zero is considered to be more temporally synchronous while a value less than zero is considered to be less temporally synchronous. In all 17 networks, our method outperforms the original atlas with significance $p < 0.005$. This highlights the importance of our subject-specific approach for glioma patients, whose functional networks are substantially reorganized due to tumor presence.

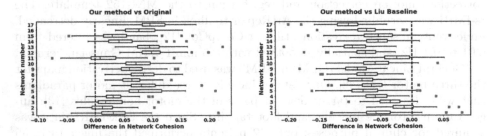

Fig. 3. Difference in intra-network cohesion between our method and the original parcellation (**left**) and the Liu baseline (**right**).

Naturally, the Liu baseline achieves higher NC due to its correlation-based voxel reassignment procedure. Figure 4 shows the original parcellation, and the final network assignment using our method and Liu's method in a single patient. Each distinct color represents one of the 17 networks. We observe an overall lack of spatial contiguity in the Liu baseline, as highlighted in the white circle. This might be due to spurious noise within rs-fMRI signal at the voxel level, resulting in some spatially discontiguous reassignment. The large grey area in the right hemisphere is the excluded tumorous region for this subject.

Figure 5 shows the proportion of voxels retained in the original network membership between our method (left) and the Liu baseline (right). We observe substantial reorganization in the networks defined from our method. Along with higher NC, this further motivates our approach, showing that many voxels in the original parcellation may not belong to the proper RSN for this cohort. We observe an even larger reorganization in the Liu networks. In the following section we conjecture that the displacement in the Liu networks may be too large, because while the Liu baseline provides more temporally cohesive RSNs, it fails to identify functionally consistent motor networks.

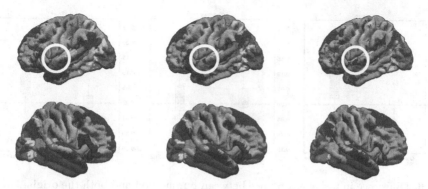

Fig. 4. Left: Original network assignment. **Middle:** Our final network assignment. **Right:** Liu's final network assignment. For visualization, we have dilated the networks according to the liberal Yeo mask.

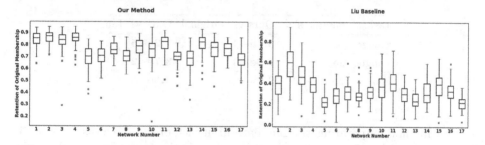

Fig. 5. Network retention for our method (**left**) and Liu's method (**right**).

3.2 Motor Network Concordance, as Validated by Task-fMRI

Our second experiment quantifies the rs-fMRI concordance between the pseudo-ground truth motor network in each patient and the motor RSN identified by each of the methods. Specifically, we will use the GLM activation map across three distinct motor tasks to define seed locations for motor functionality. The seed is defined as a group of highly activated voxels within the activation map. The Yeo atlas separates the motor network into two different parcels [9]. Our measure of task concordance will be the maximum correlation between the reference signals of these two RSNs and the average time series associated with the GLM activation seed. We determine that a method is better at motor network identification by having a higher positive correlation with significance $p < 0.05$.

Figure 6 illustrates the performance gain of our method. The pink boxplots show the difference in task concordance between our method and the original atlas, while the blue shows the difference in task concordance between our method and the Liu baseline. The tasks are ordered as finger, tongue, and foot from left to right. Table 1 summarizes the results and corresponding p-values for this experiment. The values in bold show when our method outperforms other methods with a student t-test with significance threshold $\alpha = 0.05$.

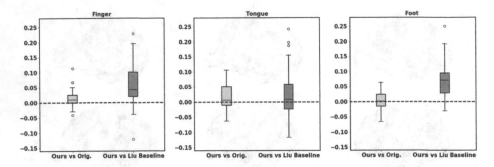

Fig. 6. Difference in task concordance between our method and both the original atlas (pink) and the Liu baseline (blue). Our method achieves significantly better performance in five out of the six comparisons. (Color figure online)

Table 1. P-values for our method vs. the original atlas and the Liu baseline.

Task	Sample size	Ours vs. Original	Ours vs. Liu
Finger	42	**3.6e−3**	**1.2e−5**
Tongue	35	**7.0e−3**	**3.7e−2**
Foot	20	0.45	**9.8e−5**

Our method outperforms the Liu method in each of the three tasks. In addition, our method performs better than the original atlas in the finger and tongue task, but not the foot task. This latter result can be due to the local area of the motor homonculus that foot activaton lies in [13] or the smaller sample size. By observing p-values reported for the finger and foot task, we conclude that no reassignment would be preferrable to the Liu baseline in this experiment. However, the Liu method RSNs were the most temporally cohesive. Though network cohesion is a desirable property for RSNs [1], we have demonstrated that higher cohesion does not always lead to a functionally consistent motor network. We conjecture that (1) Liu is too liberal in the voxel reassignment, and (2) both spatial and temporal consistency are required for RSN identification.

In summary, our method balances both spatial contiguity with temporal synchrony to help describe functional networks in patients who have undergone localized neural plasticity. We observe that our method shows more cohesive RSNs for tumor patients than a population-based functional atlas. We also determine that the motor network refined by our method is a closer representation to the actual motor network in these patients. This combination of results give us confidence in our method for characterizing RSNs in a lesional population.

4   Conclusion

We have formulated a Bayesian model that can refine a population atlas on a patient-specific basis. Our model considers both spatial contiguity as well as temporal synchrony between voxels, all while handling large and variable brain lesions. Our method outperforms established baselines for identifying a functionally consistent motor network. The use of the MRF prior along with iterative voxel reassignment shows a viable balance between properties of interest in resulting RSNs. These methodological improvements broaden the applications in which one can use rs-fMRI for analysis. We have generated a method that can be translated to other patient cohorts with anatomical brain lesions, like stroke or traumatic brain injury. Our performance in assessing RSN cohesion shows that our method captures subject-specific functional organization well, even in a pathological population. Our method outperforms both baselines in terms of motor network identification, which is an important step for preoperative planning for neurosurgical resections to avoid permanent motor network damage.

Future work with our method will involve different initial atlases. Specifically, we aim to observe how our method performs with atlases of different network numbers, and different initial size (voxel membership) of networks. Methodologically, we aim to vary the number of neighbors considered in our prior model, assigning varying weights to neighbors of different geodisic distances from the center voxel. Clinically, we aim to study reorganization of sites near the glimoa, which is known to show the most neural plasticity in these patients [3].

Acknowledgements. This work was supported by the RSNA Research & Education Foundation Carestream Health RSNA Research Scholar Grant; Contract grant number: RSCH1420.

References

1. Lee, W.H., Frangou, S.: Linking functional connectivity and dynamic properties of resting-state networks. Sci. Rep. **7**(1), 16610 (2017)
2. Sair, H.I., et al.: Presurgical brain mapping of the language network in patients with brain tumors using resting-state fMRI: comparison with task fMRI. Hum. Brain Mapp. **37**(3), 913–923 (2016)
3. Duffau, H.: Lessons from brain mapping in surgery for low-grade glioma: insights into associations between tumour and brain plasticity. Lancet Neurol. **4**(8), 476–486 (2005)
4. Blumensath, T., Jbabdi, S., et al.: Spatially constrained hierarchical parcellation of the brain with resting-state fMRI. Neuroimage **76**, 313–324 (2013)
5. Craddock, R.C., et al.: A whole brain fMRI atlas generated via spatially constrained spectral clustering. Hum. Brain Mapp. **33**(8), 1914–1928 (2012)
6. Wang, D., et al.: Parcellating cortical functional networks in individuals. Nat. Neurosci. **18**(12), 1853 (2015)
7. Venkataraman, A., et al.: Bayesian community detection in the space of group-level functional differences. IEEE Trans. Med. Imaging **35**(8), 1866–1882 (2016)
8. Ryali, S.: A parcellation scheme based on von Mises-fisher distributions and Markov random fields for segmenting brain regions using resting-state fMRI. NeuroImage **65**, 83–96 (2013)

9. Thomas, B., Yeo, B.T., et al.: The organization of the human cerebral cortex estimated by intrinsic functional connectivity. J. Neurophysiol. **106**(3), 1125–1165 (2011)
10. Wainwright, M.J., Jordan, M.I., et al.: Graphical models, exponential families, and variational inference. Found. Trends® Mach. Learn. **1**(1–2), 1–305 (2008)
11. Kschischang, F.R., Frey, B.J., Loeliger, H.-A.: Factor graphs and the sum-product algorithm. IEEE Trans. Inf. Theory **47**(2), 498–519 (2001)
12. Behzadi, Y., et al.: A component based noise correction method (compcor) for bold and perfusion based fMRI. Neuroimage **37**(1), 90–101 (2007)
13. Jack Jr., C.R., et al.: Sensory motor cortex: correlation of presurgical mapping with functional MR imaging and invasive cortical mapping. Radiology **190**(1), 85–92 (1994)

Data-Specific Feature Selection Method Identification for Most Reproducible Connectomic Feature Discovery Fingerprinting Brain States

Nicolas Georges, Islem Rekik$^{(\boxtimes)}$, and for the
Alzheimers's Disease Neuroimaging Initiative

BASIRA lab, CVIP group, School of Science and Engineering, Computing,
University of Dundee, Dundee, UK
irekik@dundee.ac.uk
http://www.basira-lab.com

Abstract. Machine learning methods present unprecedented opportunities to advance our understanding of the connectomics of brain disorders. With the proliferation of extremely high-dimensional connectomic data drawn from multiple neuroimaging sources (e.g., functional and structural MRIs), effective feature selection (FS) methods have become indispensable components for (i) disentangling brain states (e.g., early vs late mild cognitive impairment) and (ii) identifying connectional features that might serve as biomarkers for treatment. Strangely, despite the extensive work on identifying stable discriminative features using a particular FS method, the challenge of choosing the best one from a large pool of existing FS techniques for optimally achieving (i) and (ii) using a dataset of interest remains unexplored. In essence, the question that we aim to address in this work is: "Given a set of feature selection methods $\{FS_1, \ldots, FS_K\}$, and a dataset of interest, which FS method might produce the *most reproducible and 'trustworthy'* connectomic features that accurately differentiate between two brain states?" This paper is an attempt to address this question by evaluating the performance of a particular feature selection for a specific data type in fulfilling criteria (i) and (ii). To this aim, we propose to model the relationships between a set of FS methods using a multi-graph architecture, where each graph quantifies the feature reproducibility power between graph nodes at a fixed number of top ranked features. Next, we integrate the reproducibility graphs with a discrepancy graph which captures the difference in classification performance between FS methods. This allows to identify, for a dataset of interest, the 'central' node with the highest degree, which reveals the most reliable and reproducible FS method for the target brain state classification task along with the most discriminative features fingerprinting these brain states. We evaluated our method on multi-view brain connectomic data for late mild cognitive impairment vs Alzheimer's disease classification. Our experiments give insights into *reproducible* connectional features fingerprinting late dementia brain states.

© Springer Nature Switzerland AG 2018
G. Wu et al. (Eds.): CNI 2018, LNCS 11083, pp. 99–106, 2018.
https://doi.org/10.1007/978-3-030-00755-3_11

1   Introduction

Neurological and neuropsychiatric disorders, including Autism spectrum disorder (ASD), Alzheimer's disease (AD) or Mild Cognitive Impairment (MCI), are distinctive conditions that affect the morphology, cognition, and function of the brain. Understanding the connectomics of these brain disorders [1] can help improve diagnosis, prognosis, and patient treatment. To this aim, several works leveraged machine learning techniques [2–4] as well as graph analysis techniques [5] to discover distinctive brain features which reliably differentiate between normal subjects and disordered patients. These might serve as biomarkers, which can be targeted for developing efficient treatment. Due to the high dimensionality of connectomic data, many machine learning methods embed feature selection (FS) techniques to effectively reduce the dimensionality of data samples by selecting a subset of highly relevant features. Despite the great progress made over the last decade in devising robust and accurate FS methods [6], developing a new approach that would produce the best classification results and identify the most reliable feature for *all* data types seems to be an intractable problem. In fact, the ongoing proliferation of multi-source medical data, including structural and functional magnetic resonance imaging (MRI) data collected for the human brain connectome project [7], presents unprecedented challenges to devise feature selection methods that generate reproducible biomarkers across different data sources. This is because each data source has its unique characteristics and statistical distribution that might not match that of another data source. Hence, identifying the best feature selection method that unravels the inherent traits of a particular dataset remains a major challenge.

Despite the great potential that many FS methods hold for identifying connectomic biomarkers for neurological disorders (e.g., Tourette Syndrome, ASD) [8–10], training on small datasets comes with its limitations including an observable variability of most discriminative features. Being able to rely on a stable FS method that is 'optimal' for a specific dataset would constitute a radical change for detecting disordered brain changes through the connectome data. Our hypothesis is that the best performing FS method for a dataset of interest might not be optimal for a different dataset in terms of classification accuracy and feature reproducibility. To the best of our knowledge, existing FS assessment criteria have mainly focused on the stability criteria [11,12], which quantifies the sensitivity of feature selection methods to variations in the training set. However, this does not assess the suitability of the 'selected' FS method for a particular dataset. Basically, the question that we aim to address in this work is: Given a set of feature selection methods $\{FS_1, \ldots, FS_N\}$, and a particular dataset, which FS method might produce the *most reproducible and 'trustworthy'* connectomic features that accurately differentiate between two brain states (e.g., demented vs healthy)?

In contrast to methods focusing on boosting the accuracy of FS methods [13] in classifying different brain states, our primary goal is not to maximize individual-level classification accuracy but to identify the best FS method that will produce reproducible brain features associated with a specific brain disorder (i.e., potential biomarkers) for a particular dataset. To do so, given a set of FS methods, we first model the relationship between FS methods using a set of graphs, each graph quantifies the feature reproducibility power between neighboring nodes at a fixed number of top ranked features (i.e., a 'feature threshold' K). The weight of an edge connecting two FS nodes in the graph captures the overlap in top K ranked features. Next, we generate a discrepancy FS graph, where the strength of an edge connecting two FS nodes encodes the absolute difference in their classification accuracy. Ultimately, by merging all reproducibility and discrepancy graphs, we generate a holistic graph which allows the identify the central FS method with most reproducible features *in relation* to other FS methods in the graph. More importantly, the selected central FS method will be used to identify the most meaningful and reproducible connectomic features for a brain disorder of interest.

2 Multi-graph Based Identification of Data-Specific Feature Selection Method for Reproducible Discriminative Feature Discovery

In this section, we introduce the proposed pipeline to identify the FS method that produces 'the most agreed upon' features for distinguishing between two groups drawn from a connectomic data of interest. Fig. 1 displays the key steps of our framework.

Multi-view Connectomic Feature Extraction. Each brain is represented by a set of n_v networks $\{\mathbf{V}_i\}_{i=1}^{n_v}$, each encoding a particular view of the connectional brain construct. To train our classification model based on the identified FS method, we define a feature vector $\mathbf{v}_k$ for each brain network view k, whose elements belong to the off-diagonal upper triangular part of the corresponding connectivity matrix (Fig. 1).

FS-to-FS Multi-graph Construction. Given a particular data view, we aim to identify the best feature selection method that gives the most reproducible and reliable features allowing to tease apart two brain states. We hypothesize that the most reliable FS method is able to reproduce the majority discriminative features identified by other methods, thereby achieving the highest consensus with other FS methods. The most appealing characteristic of the approach is that it evaluates the importance of a given FS method while considering a set of FS methods at a given cut-off threshold K representing the number of top K ranked features selected to train the classifier (e.g., support vector machine –SVM). Given a set of N FS methods $\mathcal{F} = \{FS_1, \ldots, FS_N\}$, we construct an undirected fully-connected graph $G_K = (V_K, E_K)$; V_K is the set of nodes, each nesting an FS method in $\mathcal{F}$, while E_K represents weighted edges, which model pairwise overlap in top K features among FS methods. By varying the cut-off values K, we define a set of graphs $\mathcal{G}$ (or multi-graph) that model the overlap between FS methods at different levels. Next, for easily merging the generated multiple graphs, we represent each G_K as a similarity matrix $\mathbf{S}_K$ (Fig. 1), where each element $\mathbf{S}_K(i, j)$ denotes the overlap in top K ranked features between

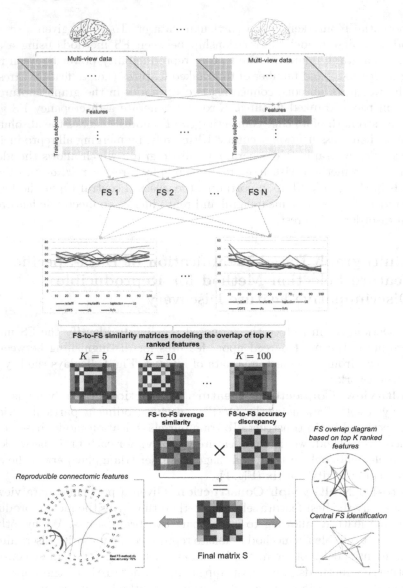

Fig. 1. *Proposed data-specific feature selection method identification pipeline.* For each subject, we define connectomic feature vectors, each derived from a particular brain view. We note that the performance of different FS methods varies with data types. Given a particular data view, we define multiple graphs, each represented as a similarity matrix modeling the consensus in top K ranked features among other selection methods. Next, we define a accuracy discrepancy matrix measuring the pairwise absolute difference in average accuracy between FS methods. By merging consensus similarity defined at multiple thresholds K with the accuracy discrepancy matrix, we generate a final matrix **S**. The best FS method for the dataset of interest is defined as the node with the highest centrality in **S**, thereby allowing to identify the most reproducible features distinguishing between two brain states (e.g., healthy vs disordered states).

FS methods i and j. We generate an average similarity matrix $\bar{S}$ by merging all similarity matrices across all thresholds, thereby capturing the *average FS method consensus* with other methods.

FS-to-FS Accuracy Discrepancy Matrix Construction. Since classification accuracy influences the credibility of the produced distinctive features, we propose to model the relationship between FS methods in terms of discrepancy in average classification accuracy. Hence, we define an average accuracy discrepancy matrix $\bar{A}$, where the cost $\bar{A}(i, j)$ of an edge connecting two nodes i and j is defined as $\bar{A}(i, j) = |\bar{a}_i - \bar{a}_j|$, where $\bar{a}_i$ represents the average accuracy of FS method i at different cut-off thresholds. Next, we merge both $\bar{A}$ and $\bar{S}$ to output the final FS similarity matrix S (Fig. 1).

FS Method Identification. We assign a score c_i for each FS_i in S, that quantifies the consensus in top selected feature set as well as classification performance among other methods. In particular, inspired from graph analysis theory, we define c_i as the centrality measure, indicating the number of times that FS method is visited on whatever path of a given length. The final FS method is selected as the one with the highest centrality in S. Once, we identify the most reliable FS method, we train an SVM classifier using the top K selected features by FS to reveal the most discriminative ones.

3   Results and Discussion

Evaluation Dataset. To perform the classification, we used leave-one-out cross validation on 77 subjects (41 AD and 36 late MCI) from ADNI data[1], each with structural T1-w MR image [14]. We reconstructed both right and left cortical hemispheres for each subject from T1-w MRI using FreeSurfer software [15]. Then we parcellated each cortical hemisphere into 35 cortical regions using Desikan-Killiany Atlas [15]. We generated two brain network datasets derived from the maximum principal curvature brain view and the mean cortical thickness brain view, respectively. For each cortical attribute, we produced a morphological brain network, where the strength of a connection linking ROI i to ROI j is defined as the absolute difference between the averaged attribute values in both ROIs [2,3].

FS Methods and Training. We used the Feature Selection Library [16] provided by Matlab to select 7 FS methods: relieff [17], mutinffs [18], laplacian [19], L0 [20], UDFS [21], llcfs [22], and cfs [23]. We adopted a leave-one-out cross-validation (CV) strategy to train each FS in combination with an SVM classifier. For FS methods that required parameter tuning, we used nested CV. For each FS method, we evaluated the performance of SVM classifier across different number of top K selected features varying from 10 to 100, with a step size of 10 features.

Findings and Future Improvements. Fig. 2 confirms our hypothesis that the best FS method for one data type might not be the best for another data type.

---

[1] http://adni.loni.usc.edu/.

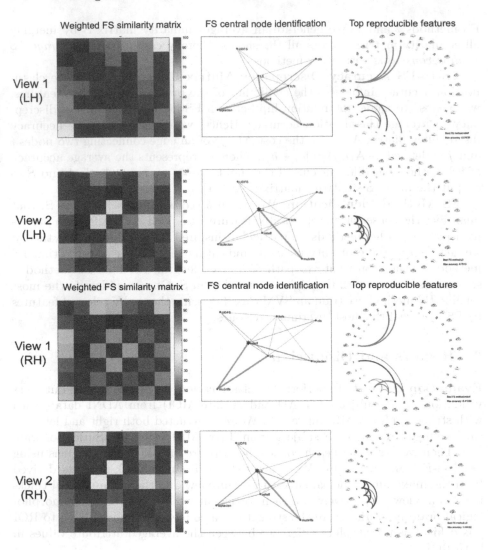

Fig. 2. *Top 10 reproducible discriminative features identification using the best identified feature selection (FS) method for each network brain view data.* Selected FS methods (⋆), corresponding classification accuracy, and top reproducible features varied across data types and right and left hemispheres (RH and LH).

For instance, relieff was identified for view 1 LH connectomic data with a classification accuracy of 61.03%, while L0 was identified for view 2 LH connectomic data with a classification accuracy of 70.3%. Overall, the identified discriminative features distinguishing between LMCI and AD brain states varied across views and cortical hemispheres. However, we note that nodes 1, 2 and 5 corresponding with the bank of the superior temporal sulcus, caudal anterior-cingulate cortex, and cuneus cortex were frequently selected. These regions were reported in other studies on AD [24].

There are several future directions to explore to further improve our seminal work. First, instead of pre-defining a similarity matrix modeling the relationship between FS methods in terms of top ranked feature consensus, we can instead learn these associations in a more generic way. Second, we will integrate the feature stability criteria for FS method identification. Third, we will evaluate our method on multiple connectomic datasets, including functional and structural connectomes. Fourth, ideally, the FS method giving the best classification accuracy would identify the most discriminative and reproducible features. We aim to further improve our framework to identify the data-specific FS method that satisfies both criteria.

4 Conclusion

In this work, we investigated a novel problem arising from the need to discover the most reproducible and reliable clinical biomarkers that distinguish between two groups (e.g., healthy and disorders brains) by identifying the best feature selection method suited for the dataset of interest. We first proposed the concept of FS similarity multi-graph to model the relationships between different FS methods in terms of overlap top ranked features at multiple thresholds. By further integrating an accuracy discrepancy graph with the similarity multigraph to enforce a consistency between high classification performance and feature reproducibility when identifying the best FS method for the target input data. By exploring the topological properties of the merged graph, we mark the central FS node with the highest the centrality score as the most reliable one. Our preliminary findings showed that the performance of a particular FS method to train a typical classifier varies with the data type. Besides classification accuracy, it is also possible to integrate feature stability as a measure to identify the best FS method. Another line of our ongoing work is to study the reproducibility of the identified features by the 'best' FS methods across *multi-source* medical datasets.

References

1. Fornito, A., Zalesky, A., Breakspear, M.: The connectomics of brain disorders. Nat. Rev. Neurosci. **16**, 159 (2015)
2. Mahjoub, I., Mahjoub, M.A., Rekik, I.: Brain multiplexes reveal morphological connectional biomarkers fingerprinting late brain dementia states. Sci. Rep. **8**(1), 4103 (2018)
3. Lisowska, A., Rekik, I.: Joint pairing and structured mapping of convolutional brain morphological multiplexes for early dementia diagnosis. Brain Connectivity (2018). https://doi.org/10.1089/brain.2018.0578
4. Zhao, F., Zhang, H., Rekik, I., An, Z., Shen, D.: Diagnosis of autism spectrum disorders using multi-level high-order functional networks derived from resting-state functional MRI. Front. Hum. Neurosci. **12**, 184 (2018). https://doi.org/10.3389/fnhum

5. Bullmore, E., Sporns, O.: Complex brain networks: graph theoretical analysis of structural and functional systems. Nat. Rev. Neurosci. **10**, 186 (2009)
6. Liu, H., Motoda, H.: Computational methods of feature selection (2007)
7. Van Essen, D.C., Glasser, M.F.: The human connectome project: progress and prospects. Cerebrum Dana Forum Brain Sci. **2016** (2016)
8. Lisowska, A., Rekik, I.: Pairing-based ensemble classifier learning using convolutional brain multiplexes and multi-view brain networks for early dementia diagnosis. In: Wu, G., Laurienti, P., Bonilha, L., Munsell, B.C. (eds.) CNI 2017. LNCS, vol. 10511, pp. 42–50. Springer, Cham (2017). https://doi.org/10.1007/978-3-319-67159-8_6
9. Soussia, M., Rekik, I.: High order connectomic manifold learning for autistic brain state identification. In: Wu, G., Laurienti, P., Bonilha, L., Munsell, B.C. (eds.) CNI 2017. LNCS, vol. 10511, pp. 51–59. Springer, Cham (2017). https://doi.org/10.1007/978-3-319-67159-8_7
10. Wen, H., et al.: Combining disrupted and discriminative topological properties of functional connectivity networks as neuroimaging biomarkers for accurate diagnosis of early tourette syndrome children. Mol. Neurobiol. **55**, 3251–3269 (2018)
11. Kalousis, A., Prados, J., Hilario, M.: Stability of feature selection algorithms: a study on high-dimensional spaces. Knowl. Inf. Syst. **12**, 95–116 (2007)
12. Lustgarten, J.L., Gopalakrishnan, V., Visweswaran, S.: Measuring stability of feature selection in biomedical datasets. In: AMIA Annual Symposium Proceedings, vol. 2009, p. 406. American Medical Informatics Association (2009)
13. Chen, Y.W., Lin, C.J.: Combining SVMs with various feature selection strategies. In: Feature extraction, pp. 315–324. Springer, Heidelberg (2006). https://doi.org/10.1007/978-3-540-35488-8_13
14. Mueller, S.G.: The alzheimer's disease neuroimaging initiative. Neuroimaging Clin. North Am. **10**, 869–877 (2005)
15. Fischl, B.: Automatically parcellating the human cerebral cortex. Cereb. Cortex **14**, 11–22 (2004)
16. Roffo, G.: Feature selection library (MATLAB toolbox). arXiv preprint arXiv:1607.01327 (2016)
17. Kononenko, I., Šimec, E., Robnik-Šikonja, M.: Overcoming the myopia of inductive learning algorithms with RELIEFF. Appl. Intell. **7**, 39–55 (1997)
18. Estévez, P.A., Tesmer, M., Perez, C.A., Zurada, J.M.: Normalized mutual information feature selection. IEEE Trans. Neural Netw. **20**, 189–201 (2009)
19. He, X., Cai, D., Niyogi, P.: Laplacian score for feature selection. In: Advances in Neural Information Processing Systems, pp. 507–514 (2006)
20. Han, J., Sun, Z., Hao, H.: l0-norm based structural sparse least square regression for feature selection. Pattern Recogn. **48**, 3927–3940 (2015)
21. Yang, Y., Shen, H.T., Ma, Z., Huang, Z., Zhou, X.: l2, 1-norm regularized discriminative feature selection for unsupervised learning. In: IJCAI Proceedings-International Joint Conference on Artificial Intelligence, vol. 22, p. 1589 (2011)
22. Zeng, H., Cheung, Y.m.: Feature selection and kernel learning for local learning-based clustering. IEEE Trans. Patt. Anal. Mach. Intell. **33**, 1532–1547 (2011)
23. Hall, M.A.: Correlation-based feature selection for machine learning (1999)
24. Wee, C.Y., Yap, P.T., Shen, D., Initiative, A.D.N.: Prediction of Alzheimer's disease and mild cognitive impairment using cortical morphological patterns. Hum. Brain Mapp. **34**, 3411–3425 (2013)

Towards Effective Functional Connectome Fingerprinting

Kendrick Li and Gowtham Atluri[✉]

Department of EECS, University of Cincinnati, Cincinnati, OH 45221, USA
likt@mail.uc.edu, atlurigm@ucmail.uc.edu

Abstract. The ability to uniquely characterize individual subjects based on their functional connectome (FC) is a key requirement for progress towards *precision neuroscience*. The recent availability of dense scans from individuals has enabled the neuroscience community to investigate the possibility of individual characterization. FC fingerprinting is a new and emerging problem where the goal is to uniquely characterize individual subjects based on FC. Recent studies reported near 100% accuracy suggesting that unique characterization of individuals is an accomplished task. However, there are multiple key aspects of the problem that are yet to be investigated. Specifically, (i) the impact of the number of subjects on fingerprinting performance needs to be studied, (ii) the impact of granularity of parcellation used to construct FC needs to be quantified, (iii) approaches to separate subject-specific information from generic information in the FC are yet to be explored. In this study, we investigated these three directions using publicly available resting-state functional magnetic resonance imaging data from the Human Connectome Project. Our results suggest that fingerprinting performance deteriorates with increase in the number of subjects and with the decrease in the granularity of parcellation. We also found that FC profiles of a small number of regions at high granularity capture subject-specific information needed for effective fingerprinting.

Keywords: Functional connectivity · Fingerprinting · Parcellation Precision neuroscience

1 Introduction

Resting state functional connectivity (RSFC) studies that estimate connectivity based on blood-oxygen-level-dependent (BOLD) signal measured using functional magnetic resonance imaging (fMRI) have revealed many principles of brain function [4,5]. Most of the existing studies made inferences about RSFC at a group level, by co-registering individual scans to a standard template, and found that such inferences are reliable [11]. While group-level inferences inform us of the generic principles, they obscure principles specific to individual subjects that are essential for characterizing brain function in health and disease. Recent availability of 'dense' fMRI scans from individuals (e.g., Human Connectome Project

© Springer Nature Switzerland AG 2018
G. Wu et al. (Eds.): CNI 2018, LNCS 11083, pp. 107–116, 2018.
https://doi.org/10.1007/978-3-030-00755-3_12

(HCP) data [12], Midnight Scan Club (MSC) [7], and MyConnectome dataset [8]) provide a tremendous opportunity to study idiosyncratic properties of brain function and make progress towards 'precision neuroscience' [10].

Functional connectome fingerprinting, where the goal is to identify individuals using subject-specific RSFC, has been explored using the above datasets that constitute dense scans from individuals [6,9]. Specifically, given a set of N *reference* fMRI scans, one from each of the N subjects, and a new *target* fMRI scan from one of the same N subjects, the goal is to identify the subject by 'matching' RSFC of the target scan with that of the reference scans. As RSFC is used to match the reference and the target scans, we refer to it as a functional fingerprint. There are different approaches to using RSFC and their effect on the accuracy of fingerprinting has been studied. For instance, Finn et al. [6], using 126 subjects from HCP, reported a fingerprinting accuracy in the range of 92%–94% while using whole-brain RSFC and 98–99% using a frontoparietal-based RSFC. In another study, using 100 unrelated subjects from HCP, Amico and Goni [3] observed that by performing principal component analysis (PCA) on whole-brain RSFC and using the resultant principal components for matching, the accuracy increased from 94% to 98%. Xu et al. [15] studied the reliability of boundaries drawn between functional areas delineated using spatial gradients (the approach is discussed elaborately in [14]) and reported success rate of up to 99% using 30 subjects.

These near 100% success rates may lead one to conclude that fingerprinting is not only a relatively easy problem, but also a solved problem with no room for progress. However, this is far from reality. Note that the underlying hypothesis that drives the fingerprinting methodology is that RSFC instances from the same subject lie in close proximity, segregated from other subjects' RSFC, in some high-dimensional space. When a small number of subjects are sampled from a population, the RSFCs from one subject may be well separated from that of others in the high-dimensional space. However, when many more subjects are sampled from a population, this high-dimensional space may become cluttered with RSFCs from different subjects, where RSFCs from different subjects may look more similar than the RSFCs from the same subject, and as a result hurt the overall fingerprinting performance. This aspect of fingerprinting is yet to be studied.

In addition, the impact of granularity of the parcellation used for computing RSFC on fingerprinting accuracy is yet to be investigated. A parcellation of the brain is expected to capture functionally distinct areas at a given level of granularity, often indicated as the number of parcels. While *subject-specific* RSFC is desired for fingerprinting purpose, the granularity at which this subject-specific information becomes available is not known.

Subject-level RSFC contains both generic and subject-specific information. Separating out subject-specific information from generic information is crucial for determining what aspects of RSFC are relevant for fingerprinting. Finn et al.'s approach [6] of using RSFC within different groups of brain regions is one approach. Their underlying hypothesis is that the subject-specific signatures are present within the RSFC of different region groups. One hypothesis, that is not yet explored, is that an RSFC profile of one or small number of regions could be used for fingerprinting. This direction allows us to study the degree of subject-specific information available in a single-node's FC profile and it also allows us to discover the regions in the brain that provide subject-specific connectivity maps for fingerprinting.

In this study, we investigated the above directions to deepen our understanding of functional connectome fingerprinting. Specifically, we addressed the following three questions: (1) How does the number of subjects affect the accuracy of fingerprinting? (2) How does the granularity of parcellation used for computing RSFC affect fingerprinting accuracy? (3) Can we find RSFC elements that are highly suited for effective fingerprinting? We performed our analysis on resting state fMRI data from 339 unrelated individuals in the HCP, using computing resources from the Ohio Supercomputer Center [16]. Our results suggest that fingerprinting performance deteriorates with increase in the number of subjects and with the decrease in the granularity of parcellation. We also found that a small number of regions at high granularity capture subject-specific information needed for effective fingerprinting.

The rest of this paper is organized as follows: The datasets used in our study are described in Sect. 2. Methods we used to answer above questions are presented in Sect. 3. We discussed our results in Sect. 4 and we concluded with Sect. 5.

2   Data

Resting state fMRI data from the 1200-subjects 2017 HCP data release (March 2017) [12] was used in this study. This release included processed resting state fMRI scans from 1003 healthy young adults. While we could use all of the 1003 subjects' data, any familial relationships among subjects may muddle our analysis for fingerprinting. To avoid familial relationships among subjects, we used a set of 339 unrelated subjects provided in the HCP release [2].

As part of the HCP, resting-state fMRI scans were collected from each subject on two separate days. On each day, a 20 min scan left-to-right (LR) phase encoded scan and a 20 min right-left (RL) phase encoded scan were obtained. For these four fMRI scans, we used the extensively-preprocessed node-timeseries data that was made available in the HCP data release. This node-timeseries data was generated by performing a series of steps including preprocessing, artefact removal using ICA, inter-subject registration, group-PCA, group-Independent Component Analysis (ICA), and dual-regression to compute time series for each independent component (IC). These steps are described in the HCP documentation [1]. As part of the Group-ICA step of the HCP preprocessing pipeline, the brain was parcellated into ICs at different granularities: 15, 25, 50, 100, 200, and 300 regions. Node-timeseries for ICs from each of these parcellations were provided in the HCP data release. We refer to the set of node timeseries from these ICs as IC_{15}, IC_{25}, IC_{50}, IC_{100}, IC_{200}, and IC_{300}. The node-timeseries data from the March 2017 release was used as is without further processing.

3   Methods

3.1   FC Fingerprinting

We will formally establish the terminology that will be used in the rest of the paper. We refer to fMRI scans for which we know which subject they are collected from as 'reference' scans. We refer to the new set of scans for which the subject they are collected from needs to be determined by matching with reference scans as 'target' scans. Given a set of N reference scans $\{R_1, R_2, \ldots, R_N\}$ from N different subjects, and a set of target scans $\{T_1, T_2, \ldots, T_N\}$ from the same set of subjects, the problem of FC fingerprinting is to determine for each target scan T_i the corresponding subject's reference scan R_j by matching their RSFC. There are two key steps here: (1) computing FC, (2) matching FC.

For computing RSFC from an IC node-timeseries derived from a scan, we computed Pearson correlation between each pair of node-timeseries. As the scans were collected from each subject on two separate days, we computed the average RSFC per day. That is, we averaged the RSFC from the resting-state LR and RL encoding scans on each day. As a result we have two RSFCs, one per day, from each subject: $RSFC_{d1}$ and $RSFC_{d2}$. As the node-timeseries data is available for different granularities of parcellation, these two RSFCs were computed for each of the granularities.

For matching RSFCs, we used a method that is similar to that of Finn et al.'s [6] whole-brain approach. Specifically, for each RSFC computed from a target scan T_i, we computed the Pearson correlation between the vector constructed by taking the upper-triangular values of the target RSFC matrix with that of each of the reference RSFCs. The reference RSFC that showed highest correlation with the target RSFC is treated as a match.

The accuracy of fingerprinting is computed as the fraction of subjects for which the target scans were perfectly matched with their reference scans. As we have two RSFCs from each subject ($RSFC_{d1}$ and $RSFC_{d2}$), we computed fingerprinting accuracy in two ways: (1) using $RSFC_{d1}$ as a reference and $RSFC_{d2}$ as target, (2) using $RSFC_{d2}$ as a reference and $RSFC_{d1}$ as target. The results from the former and latter cases are labelled as *Day1 Ref: Day2 Tgt* and *Day 2 Ref: Day1 Tgt*, respectively.

3.2   Studying the Effect of Sample Size

To study the effect of sample size on fingerprinting accuracy, we conducted fingerprinting analysis on smaller subsets of the dataset. Out of the 339 subjects in our dataset, we randomly selected samples of different sizes ($\{50, 95, 140, 185, 230, 275, 320\}$) and computed their fingerprinting accuracies using the method described above. This was repeated 100 times for each size and the average accuracies for the 100 runs are reported.

Silhouette Coefficient Based Analysis: To investigate the effect of the sample size further and to test our hypothesis that with more and more subjects the RSFC space gets cluttered making it difficult to perform fingerprinting accurately, we used Silhouette coefficient [13], a commonly used cluster evaluation metric, to determine how well separated the subjects' RSFCs are in the space. A Silhouette value is computed for each data point in the cluster and the value can only range from −1 to 1. Positive values closer to 1 indicate that the data point is at the core of the cluster, while a value closer to −1 indicates that the point is actually closer to points in another cluster than in the same cluster. For our analysis, an RSFC with a negative value is indicative that it is more similar to RSFCs from other subjects than it is to the RSFCs from the same subject. For a more complete treatment of Silhouette coefficient we refer an interested reader to [13]. The Silhouette value was calculated for each RSFC by assigning all RSFCs from each subject to a separate cluster. For this analysis, we used RSFCs computed from all four scans of a subject and so each cluster has four members. We computed the average Silhouette value over all RSFCs from all subjects. To understand how sample size affects the space of RSFCs using Silhouette coefficient, we created 100 randomly sampled sets of subjects each for different sample-sizes ($\{5, 50, 95, 140, 185, 230, 275, 320\}$). For each sample-size, we computed the average Silhouette coefficient for each of the hundred sets. We also computed the fraction of subjects that contained a scan with a negative Silhouette value in each of the sets for different sample-sizes. We reported the average of the fraction of subjects.

3.3   Studying the Effect of Granularity of Parcellation

To study the effect of the granularity of parcellation on fingerprinting accuracy, we performed fingerprinting analysis on 100 randomly sampled subjects using their node-timeseries from IC_{15}, IC_{25}, IC_{50}, IC_{100}, IC_{200}, and IC_{300}. For each level of granularity, the fingerprinting accuracy was recorded. This was repeated 100 times and the average accuracy for each granularity are reported.

3.4   Determining Elements of RSFC that are Highly Relevant for Fingerprinting

Subject-level RSFC contains both generic and subject-specific information. Separating out subject-specific information from generic information is crucial for determining what elements of RSFC are relevant for fingerprinting. We pursue two key methods for identifying relevant RSFC components. Our approach is to select the RSFC profile for one brain region, i.e., all region-pairs that involve the brain region, and compute the fingerprinting performance of that region's FC profile.

Single-Node RSFC Based Fingerprinting. The FC fingerprinting method described above (in Sect. 3.1) uses the entire set of elements from an RSFC. Our hypothesis is that only one region's connectivity profile may be sufficient to uniquely fingerprint a subject. To test this, we use only edges incident on one

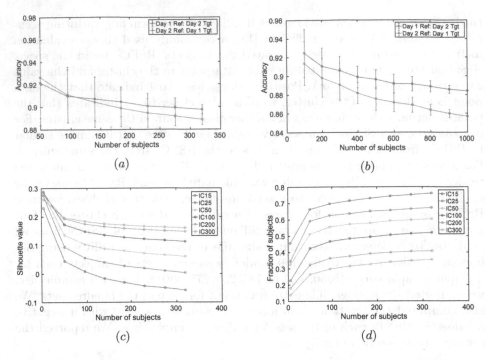

Fig. 1. The effect of the number of subjects on RSFC fingerprinting. (a) Average accuracy of fingerprinting as the number of subjects increased from 50 to 320 using the IC_{300} dataset. The error bars indicate accuracies one standard deviation away from the mean. (b) Accuracy of fingerprinting on the 1000 subject IC_{300} dataset as the number of subjects increased from 100 to 1000. Error bars indicate accuracies one standard deviation away from the mean. (c) The average subject Silhouette values with varying number of subjects. (d) The fraction of subjects with a negative Silhouette value for at least one RSFC.

region (at a time) for matching a target RSFC with reference RSFCs. This is repeated for all the regions in the parcellation to determine the regions whose RSFC profile captures highly subject-specific information. We randomly selected 100 subjects and conducted the fingerprinting analysis for each parcellation granularity on each node and recorded the resultant accuracies.

Studying Reliability of Single Node Analysis. We randomly selected a set of 150 subjects from the 339 unrelated individuals and we refer to it as 'Group A'. From the remaining individuals, we randomly selected another set of 150 subjects and refer to it as 'Group B'. Using RSFCs computed from IC_{300} dataset, we computed for each of the 300 nodes their FC fingerprinting accuracy when single node RSFC is used for groups A and B separately. We compare these node-level fingerprinting accuracies from groups A and B to determine the reliability of the single node fingerprinting accuracies.

4   Results

4.1   The Effect of Sample Size on Fingerprinting

The results from our analysis of quantifying the effect of sample size on fingerprinting are shown in Fig. 1. The fingerprinting accuracy for unrelated individuals decreased from 92.16% to 89.75% as the number of subjects increased from 50 to 320 for the scenario *Day 1 Ref: Day 2 Tgt*. Similar reduction in accuracies were seen for *Day 2 Ref: Day 1 Tgt*, even though these accuracies are relatively small compared to *Day 1 Ref: Day 2 Tgt* (Fig. 1(a)). Note that the observed (2–3%) decrease in fingerprinting accuracy (in Fig. 1(a)) may seem tolerable, but when FC fingerprinting is considered for

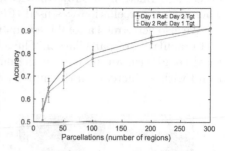

Fig. 2. The accuracy of fingerprinting with change in parcellation granularity. The error bars indicate accuracies one standard deviation away from the mean.

clinical practice where the underlying sample size is significantly larger the estimated accuracy may not meet the demands of precision neuroscience. To understand the extent of this drop in accuracy on larger datasets, we performed this analysis on the larger HCP dataset with 1000 subjects. The accuracy for 1000 subjects was 85.8%, for the scenario *Day 2 Ref: Day 1 Tgt*. These results suggest that there can be a significant reduction in accuracies as larger and larger datasets are considered.

In general, this reduction in accuracy could be due to RSFCs from different subjects exhibiting more similarity than the RSFCs from the same subject with the increase in the number of subjects. That is, the space of RSFCs is more cluttered as the number of subjects increased. To further investigate this hypothesis of cluttering in RSFC space due to increased number of subjects, we used Silhouette coefficient, a popular cluster evaluation metric, to quantify segregation of RSFCs. The average subject Silhouette value decreased from 0.2608 to 0.1606 as the number of subjects increased from 5 to 320 subjects for IC_{300} (Fig. 1(c)). This supports our hypothesis that the space of RSFCs becomes less segregated (or more cluttered) as the number of subjects increased. Furthermore, there was an increase in the fraction of subjects with a negative Silhouette value for at least one RSFC from an average value of 13.8% to 35.51% as the number of subjects increased from 5 to 320 for IC_{300} (Fig. 1(d)). This quantifies the degree of cluttering in the RSFC space as a function of sample size.

4.2   The Effect of Parcellation Granularity on Fingerprinting

We also saw an increase in fingerprinting accuracy as the granularity of parcellation increased (Fig. 2). The average accuracy increased from 55.47% to 91.13% as the number of parcels increased from 15 to 300 for the scenario *Day 1 Ref:*

Day 2 Tgt. This suggests that finer parcellations capture subject-specific RSFC more effectively than coarser parcellations. This result is also in agreement with our previous Silhouette results (Fig. 1(c)); in all cases the Silhouette values were lower, and fraction of subjects with a negative Silhouette value were higher, when coarse parcellation was used (Fig. 1(c) and (d)).

We also performed a combined analysis on the effect of the number of subjects and granularity of parcellation. The average accuracy showed a constant downward trend with an increase in the number of subjects and a constant upward trend with an increase in the number of parcels (Fig. 3).

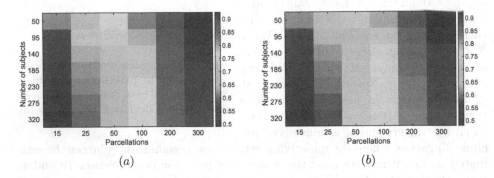

(a)   (b)

Fig. 3. Heatmap showing the relation between the number of subjects and the granularity of parcellation on fingerprinting accuracy. (a) *Day 1 Ref: Day 2 Tgt* (b) *Day 2 Ref: Day 1 Tgt*.

4.3   Determining Elements of RSFC that Are Highly Relevant for Fingerprinting

We computed fingerprinting accuracy for each brain region by 'matching' the edges incident on the region from the target RSFC with the reference RSFCs. This was repeated for each parcellation granularity. The results are shown in Fig. 4(a) and (b). There are three key observations: (1) There is an increase in the range of fingerprinting accuracy as the number of nodes increased (Fig. 4(a) and (b)). (2) The best single-node accuracy for finer parcellations are nearly as good as the whole-brain RSFC based accuracy. For instance, best single-node accuracy for IC_{300} was 86.13% compared to the whole-brain RSFC accuracy of 91.13% (only 5% lower) for *Day 1 Ref: Day 2 Tgt*. (3) The difference between the best single-node accuracy and the whole-brain RSFC decreased with increase in the number of parcellations. These results suggest that at a finer granularity of parcellation, some region's RSFC not only captures subject-specific information but also does so nearly as well as the whole-brain RSFC.

To assess the reliability of the accuracies across two different samples of subjects, we created two non-overlapping groups A and B of 150 subjects each and computed single node RSFC based accuracies separately. The single-node

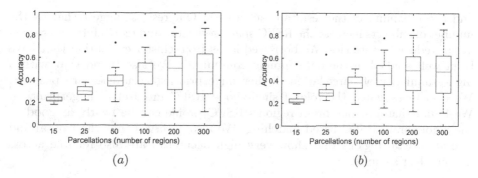

Fig. 4. Single node RSFC based fingerprinting accuracy: (a) *Day 1 Ref: Day 2 Tgt* (b) *Day 2 Ref: Day 1 Tgt*. The accuracy of using the full RSFC for fingerprinting as a black dot for each parcellation granularity.

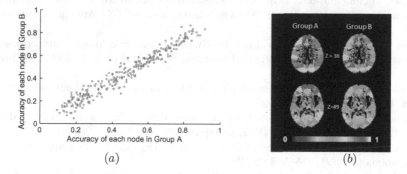

Fig. 5. (a) Comparision between the single-node RSFC-based fingerprinting accuracy between groups A and B for *Day 1 Ref: Day 2 Tgt*. (b) The accuracies of each node are colored in the brain volume for groups A and B.

accuracies of 300 nodes in the IC_{300} parcellation are strongly correlated between groups A and B (Fig. 5(a)). This suggests that these regions that consistently resulted in higher accuracies in independent samples capture FC information unique to individual subjects. The fingerprinting accuracies for the components in the IC_{300} dataset for groups A and B are shown in Fig. 5(b). The regions that resulted in higher accuracies in group A also resulted in higher accuracies in group B. Particularly, the ICs in the frontal region and lateral-parietal regions resulted in highest accuracy, approximately 90%, among other regions. These results are consistent with the findings reported in Finn et al. [6], where they observed frontoparietal network to exhibit very high accuracy.

5   Conclusion

In this work we investigated the different aspects of FC fingerprinting that have been overlooked. They include the impact of number of subjects and granularity of parcellation. We also studied single-node RSFC-based fingerprinting and the reliability of the resultant accuracies. Our results suggest that as the number of subjects increase the RSFC space gets more and more cluttered resulting in reduced accuracies. We borrowed ideas from cluster evaluation that have been well studied in the data mining community. We also found that with a high-granularity of parcellation, higher fingerprinting accuracies are possible. We also investigated the role of single-node RSFC in effective fingerprinting. We found that just one brain region's RSFC profile can be nearly as good as the whole-brain RSFC based matching. We also observed that the frontal and lateral-parietal regions that show very high accuracies are also reliable across independent samples.

References

1. HCP documentation. https://www.humanconnectome.org/storage/app/media/documentation/s1200/HCP1200-DenseConnectome+PTN+Appendix-July2017.pdf
2. List of 339 unrelated individuals in HCP. https://wiki.humanconnectome.org/download/attachments/89391484/Unrelated_S900_Subject_multilist1_with_physio.csv
3. Amico, E., Goñi, J.: The quest for identifiability in human functional connectomes. Sci. Rep. **8**(1), 8254 (2018)
4. Atluri, G., et al.: The brain-network paradigm: using functional imaging data to study how the brain works. IEEE Computer **49**(10), 65–71 (2016)
5. Bandettini, P.A.: Twenty years of functional MRI: the science and the stories. Neuroimage **62**(2), 575–588 (2012)
6. Finn, E.S., et al.: Functional connectome fingerprinting: identifying individuals using patterns of brain connectivity. Nat. Neurosci. **18**(11), 1664 (2015)
7. Gordon, E.M., et al.: Precision functional mapping of individual human brains. Neuron **95**(4), 791–807 (2017)
8. Laumann, T.O., et al.: Functional system and areal organization of a highly sampled individual human brain. Neuron **87**(3), 657–670 (2015)
9. Miranda-Dominguez, O., et al.: Connectotyping: model based fingerprinting of the functional connectome. PloS one **9**(11), e111048 (2014)
10. Poldrack, R.A.: Precision neuroscience: dense sampling of individual brains. Neuron **95**(4), 727–729 (2017)
11. Shehzad, Z., et al.: The resting brain: unconstrained yet reliable. C.Cortex (2009)
12. Smith, S.M.: Resting-state fMRI in the human connectome project. Neuroimage **80**, 144–168 (2013)
13. Tan, P.N., Steinbach, M., Kumar, V.: Introduction to Data Mining (2006)
14. Wig, G.S.: An approach for parcellating human cortical areas using resting-state correlations. Neuroimage **93**, 276–291 (2014)
15. Xu, T.: Assessing variations in areal organization for the intrinsic brain: from fingerprints to reliability. Cereb. Cortex **26**(11), 4192–4211 (2016)
16. Ohio Supercomputer Center (1987). http://osc.edu/ark:/19495/f5s1ph73

Connectivity-Driven Brain Parcellation
via Consensus Clustering

Anvar Kurmukov[1,2], Ayagoz Musabaeva[1], Yulia Denisova[1], Daniel Moyer[4],
and Boris Gutman[1,3(✉)]

[1] The Institute for Information Transmission Problems, Moscow, Russia
[2] National Research University Higher School of Economics, Moscow, Russia
[3] Illinois Institute of Technology, Chicago, USA
bgutman1@iit.edu
[4] University of Southern California, Los Angeles, USA

Abstract. We present two related methods for deriving connectivity-based brain atlases from individual connectomes. The proposed methods exploit a previously proposed dense connectivity representation, termed continuous connectivity, by first performing graph-based hierarchical clustering of individual brains, and subsequently aggregating the individual parcellations into a consensus parcellation. The search for consensus minimizes the sum of cluster membership distances, effectively estimating a pseudo-Karcher mean of individual parcellations. We assess the quality of our parcellations using (1) Kullback-Liebler and Jensen-Shannon divergence with respect to the dense connectome representation, (2) inter-hemispheric symmetry, and (3) performance of the simplified connectome in a biological sex classification task. We find that the parcellation based-atlas computed using a greedy search at a hierarchical depth 3 outperforms all other parcellation-based atlases as well as the standard Dessikan-Killiany anatomical atlas in all three assessments.

1 Introduction

The ability to quantify how the human brain is interconnected in vivo has opened the door to a number of possible analyses. In nearly all of these, brain parcellation plays a crucial role. Variations in parcellation significantly impact connectome reproducibility, derived graph-theoretical measures, and the relevance of connectome measures with respect to biological questions of interest [16]. A natural approach is then to use individual densely sampled connectomes to drive the parcellation directly, leading to a more compact, connectivity-aware set of brain regions and resulting graph, as done in e.g. [10]. A comprehensive review of parcellation methods and their effects on the derived connectome quality is given in [17]. Because individual connectivity data is at once very informative and highly redundant, there is a great flexibility in how parcels can be derived from dense, highly resolute graphs. It is possible for example to derive (1) a unified population-based atlas, (2) individual-level parcellations with cross-subject

© Springer Nature Switzerland AG 2018
G. Wu et al. (Eds.): CNI 2018, LNCS 11083, pp. 117–126, 2018.
https://doi.org/10.1007/978-3-030-00755-3_13

label mapping, or (3) individual parcellations with no inter-subject label correspondence. While the first approach is appealing for its simplicity and ease of interpretation, the second and third may enable the researcher to reveal some individual aspect of the connectome that is lost in the aggregate atlas.

In this work, we attempt to bridge these three approaches by first constructing maximally flexible hierarchical parcellations, and then finding a unifying set of labels and parcels to maximize individual agreement. We use the a continuous representation of a brain connectivity [8] as our initial dense connectome representation. Continuous connectivity is a parcellation-free representation of tractography-based, or "structural" connectomes that is based on the Poisson point process. Once individual parcellations are computed, we obtain a groupwise parcellation using partition ensemble algorithm. We access quality of the resulting parcellations in three ways. (1) We use the continuous connectome framework to compare parcellation-approximate and exact edge distribution functions. (2) We compare performance of the resulting graphs on a gender classification task. (3) We also show that without any explicit knowledge of brain geometry and based solely on graph connectivity we obtain comparatively symmetric parcellations.

2 Methods

2.1 Continuous Connectome

The continuous connectome model (ConCon) treats each tract as an observation of an inhomogeneous symmetric Poisson point process with the intensity function given by

$$\lambda : \Omega \times \Omega \to \mathbb{R}^+, \tag{1}$$

where Ω denote union of two disjoint toplogically spherical brain hemispheres, representing cortical white matter boundaries. In practice, ConCon uses cortical mesh vertices as nodes of connectivity graph. From such a representation, a "discrete" connectivity graph could be computed from any particular cortical parcellation P. We follow definitions from [8] and call $P = \{E_i\}_{i=1}^N$ a parcellation of Ω if $E_1 \ldots E_k \subseteq \Omega$ such that $\cup_i E_i = \Omega$, and N is the number of parcels (ROIs). Edges between regions E_i and E_j can then be computed by integration of the intensity function:

$$\mathcal{C}(E_i, E_j) = \iint_{E_i, E_j} \lambda(x, y) dx dy, \tag{2}$$

Due to properties of the Poisson Process, $\mathcal{C}(E_i, E_j)$ is the expectation of the number of observed tracts between E_i and E_j. In the context of connectomics, this is the expected edge strength.

2.2 Graph Clustering

Once we obtain all individual continuous connectomes, we partition each independently into a set of disjoint communities. For graph clustering we use the

Louvain modularity algorithm [1], as it has shown good results in multiple neuroimaging studies [5,7,9,12]. This algorithm consist of two steps. The first step
combines locally connected nodes into communities, while the second step builds
new meta graph. The nodes of the meta-graph are communities from the previous step, and the edges are defined as the sum of all inter-community connections
of the new nodes. The algorithm in [1] cycles over these steps iteratively, converging when further node clustering leads to no increase in modularity. We
follow the hierarchical brain concept [7], repeating the clustering procedure iteratively. After the initial parcellation, we further cluster each individual parcel as
an independent graph. In this work, we repeat the process three times. For each
(i'th) continuous connectome this procedure yields a three-level hierarchically
embedded partition: $P_i^{I}, P_i^{II}, P_i^{III}$, (see Fig. 1).

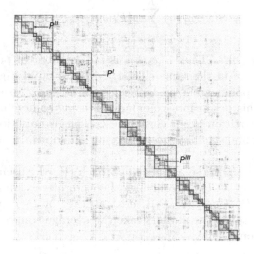

Fig. 1. Adjacency matrix of a sample continuous connectome. Rows and columns are
reordered according to partition of the third hierarchical level. Boxes of different color
represents clusters of different hierarchical levels. P^{I} clusters are obtained first, next
we reapply clustering on each detected P^{I} cluster and obtain P^{II}. This is repeated once
more to obtain P^{III} (Color figure online)

2.3   Consensus Clustering

In order to obtain a unified parcellation for all subjects, we use consensus clustering. The concept was developed for aggregating multiple partitions of the same
data into a single partition. We define the *average* partition over all individual
partitions $\{P_i\}$ as:

$$\bar{P} == \operatorname{argmin}_P \sum_i d(P, P_i), \qquad (3)$$

where $\bar{P}$ is used to denoted desirable *average* partition, K is a number of averaged partitions, $d(P_i, P_j)$ is a distance measure between two partitions and we want to minimize average distance from $\bar{P}$ to all given partitions P_i. All partitions are represented by a vector of length M, where M is a number of clustered objects (vertices of a graph in our case). It contains values from 1 up to N, where N is a number of clusters (parcels). This task is generally NP complete [14], but there are many approximate algorithms. We use two approaches: Cluster-based Similarity Partitioning Algorithm (cspa) [11] and greedy algorithm from [2].

CSPA defines a similarity between data points based on co-occurrence in a same cluster across different partitions, and then partitions a graph induced by this similarity. Specifically, given multiple partitions $P_1, \ldots P_K$ of a data points $x_1, \ldots x_M$. One can define similarity between points x_i, x_j as follow:

$$S(x_i, x_j) = \sum_{k=1}^{K} \delta(P_k(x_i), P_k(x_j)), \tag{4}$$

Here δ is Kroneker delta. Thus $S(x_i, x_j)$ is just number of partitions in which points x_i and x_j were in the same cluster. Next we build a graph, with nodes correspond to data points and edge between node x_i and x_j is equal to $S(x_i, x_j)$. We the partition this graph into communities using some clustering algorithm and the resulting partition is our clustering consensus partition.

Another way to find such *average* clustering is to optimize loss function given by Eq. 3.

The authors of [2] propose a greedy approach (Hard Ensemble - HE). Given multiple partitions $P_1 \ldots P_K$ it combines them iteratively, first it finds average of $\bar{P}_{1,2} = \min_{\bar{P}}(d(\bar{P}, P_1) + d(\bar{P}, P_2))$, next average of $P_{1,2}$ and P_3 and so on. As a measure of distance the authors take the average square distance between membership functions:

$$d(P_i, P_j) = \frac{1}{N} \sum_{k=1\ldots N} ||p_i^k - p_j^k||^2, \tag{5}$$

Exclusively for this definition we use another way to encode object's memberships: P_i is a matrix of size $M \times N$ (number of objects times number of clusters)

$$P_i^{m,n} = \begin{cases} 1 & \text{if } m\text{'th object belongs to } n\text{'th cluster} \\ 0 & \text{otherwise.} \end{cases} \tag{6}$$

In Eq. 5 p_i^k and p_j^k are k^{th} rows of memberships matrices P_i and P_j respectively. They correspond to membership vector of the kth object. Since we are looking for disjoint clusters, only a single element of such row vector is equal to 1. This representation is defined up to any column permutation π of matrix P, thus the optimization procedure is done subject to all possible column permutations.

2.4   Comparison Metrics

Once we find individual partitions and combine them into an average partition, we want to access their quality. We use two different approaches.

First, we compare representation strength of different parcellations by measuring distance between original $\lambda(x, y)$ and its piece-wise approximation given by:

$$\gamma(x, y) = \frac{1}{|E_i||E_j|} \mathcal{C}(E_i, E_j), \tag{7}$$

where $x \in E_i$ and $y \in E_j$. Natural way to compare two statistical distributions is to measure distance between their probability density functions, we will use Kullback-Leibler divergence [4]. For two probability distributions with densities $\lambda(x)$ and $\gamma(x)$ the KL divergence is:

$$KL(\lambda, \gamma) = \int_{-\infty}^{\infty} \lambda(x) \log \frac{\lambda(x)}{\gamma(x)} dx, \tag{8}$$

It takes values close to 0 if two distributions are equal almost everywhere. Similar but symmetrized version of KL divergence is Jensen-Shannon divergence [6]. Again for two probability distributions with densities $\lambda(x)$ and $\gamma(x)$ it is given by:

$$JS(\lambda, \gamma) = \frac{1}{2}(KL(\lambda, r) + KL(\gamma, r)), \tag{9}$$

where $r(x) = \frac{1}{2}(\lambda(x) + \gamma(x))$.

Second, we compare performance of different parcellations on a gender classification task. We use Logistic Regression model with (small) l_1 regularization on a vectors of edge weights (the upper triangle of adjacency matrix excluding diagonal). Classification performance is measured in terms of ROC AUC score, which is typical for binary classification tasks.

Finally, in order to quantify goodness of consensus clustering and access hemisphere symmetry we use Adjusted Mutual Information [15]. It measures similarity between two partitions, with value 1 corresponds to identical partitions and values close to zero for partitions that are very different. Given set X of n elements, $X = \{x_1, x_2, \ldots x_n\}$ let us consider two partitions of X: $U = \{U_1, U_2, \ldots U_l\}$ and $V = \{V_1, V_2, \ldots V_k\}$. These partitions are *strict* (or *hard*):

$$\bigcap_{j=1}^{k} V_j = \bigcap_{i=1}^{l} U_i = \emptyset$$

and *complete*:

$$\bigcup_{j=1}^{k} V_j = \bigcup_{i=1}^{l} U_i = X$$

We can construct the following $l \times k$ contingency table:

U, V	V_1	V_2	$\cdots$	V_k	$\sum_{j=1}^{k} s_{ij}$
U_1	s_{11}	s_{12}	$\cdots$	s_{1k}	s_1
U_2	s_{21}	s_{22}	$\cdots$	s_{2k}	s_2
$\vdots$	$\vdots$	$\vdots$	$\ddots$	$\vdots$	$\vdots$
U_l	s_{l1}	s_{l2}	$\cdots$	s_{lk}	s_l
$\sum_{i=1}^{l} s_{ij}$	s^1	s^2	$\cdots$	s^k	

Here s_{ij} denotes a number of common objects between U_i and V_j:

$$s_{ij} = \left| U_i \bigcap V_j \right|$$

then Mutual Information is given by:

$$\mathbf{MI} = \sum_{i=1}^{l} \sum_{j=1}^{k} P(i,j) \log \frac{P(i,j)}{P(i)P'(j)}, \tag{10}$$

where $P(i)$ is the probability of a random sample occurring in cluster U_i, $P'(j)$ is the probability of a random sample occurring in cluster V_j:

$$P(i) = \frac{s^i}{n}, \quad P'(j) = \frac{s_j}{n}$$

and $P(i,j)$ - probability of an object occurs in U_i and V_j simultaneously:

$$P(i,j) = \frac{s_{ij}}{n}$$

Adjustment scheme as proposed by Hubert and Arabie [3] has the following general form:

$$\text{Adjusted Index} = \frac{\text{Index} - \text{Expected Index}}{\text{Max Index} - \text{Expected Index}} \tag{11}$$

Using AMI we access ensemble goodness (how good clustering ensemble algorithm combines multiple partitions) using modified 3:

$$\text{Ensemble goodness} = \sum_{i}^{K} \text{AMI}(\bar{P}, P_i), \tag{12}$$

We compute parcellation symmetry by comparing hemisphere parcels (labels):

$$\text{Symmetry} = \text{AMI}(\bar{P}_{\mathbf{LH}}, \bar{P}_{\mathbf{RH}}). \tag{13}$$

3   Experiments

3.1   Data Description

We use construct continuous connetomes of 400 subjects from the Human Connectome Project S900 release [13] following [8]. We use an icosahedral spherical sampling, at a resolution of 10242 mesh vertices per hemisphere. We used Dipy's implementation of constrained spherical deconvolution (CSD) to perform probabilistic tractography. Prior to clustering, we exclude all mesh vertices that were labeled by FreeSurfer as corpus callosum or cerebellum.

3.2   Experimental Pipeline

Our experiments are summarized as follows:

1. For each subject we reconstruct its Continuous Connectome.
2. For each Continuous Connectome we iteratively run Louvain clustering algorithm, as described above. Subgraphs of having less then 1% of original graph vertices were not divided.
3. Next we aggregate individual subject partitions and obtain consensus clustering. Aggregation was done over 400 HCP subjects. Further, after finding the optimal parcellation, we obtain two parcellations based on two disjoint sets of 200 HCP subjects in order to compute reproducibility.
4. We aggregate partitions of the same level (I-II-III) using CSPA and HE.
5. We compare obtained partitions between themselves and with FreeSurfer's Desikan-Killiani parcellation using Kullback-Leibler and Jensen-Shannon divergence. We compute goodness of an ensemble and parcellation symmetry using AMI.
6. We compare performance of simplified connectomes on a binary classification task using Logistic Regression with l_1 penalty. Classification results are measured in terms of ROC AUC score, with averaging over 10 cross-validation folds.

3.3   Results

Table 1 represent all comparison results. First we can see that CSPA algorithm failed to find good clustering ensemble which result in poor classification performance and high KL and JS divergences. Greedy algorithm performed on P^{III} on the other hand outperforms standard Desikan atlas across all comparison metrics (except number of parcels, 68 versus 83). Surprisingly, greedy ensemble of second level partition (P^{II}) performs comparatively with Desikan, despite having twice as lower number of parcels (30 versus 68).

Another interesting property that we get automatically is parcellation symmetry. Our clustering algorithm known nothing about brain topology (all information was contained in graph connectivity), still reconstruct parcellations which

Table 1. All results are rounded to 2 significant digits. Where it possible results are reported with standard deviation. Best result in each row is colored. KL, JS divergences, **lower is better**; binary Gender Classification was measured in terms of ROC AUC score, higher - better; Ensemble goodness and Hemisphere symmetry were measured using AMI, Ensemble goodness is an average AMI between consensus partition and all individual partitions, higher - better.

	cspa P^I	cspa P^{II}	cspa P^{III}	HE P^I	HE P^{II}	HE P^{III}	DKT
KL	1.22 ± .07	1.18 ± .07	1.15 ± .07	1.16 ± .07	.86 ± .05	.66 ± .04	.83 ± .05
JS	.20 ± .00	.19 ± .00	.19 ± .00	.20 ± .00	.17 ± .00	.14 ± .00	.16 ± .00
Gender Classification	.63 ± .04	.64 ± .04	.69 ± .03	.64 ± .03	.75 ± .03	.86 ± .02	.81 ± .03
Hemisphere symmetry	.15	.24	.32	.26	.55	.66	.64
Ensemble goodness	.47 ± .06	.40 ± .02	.35 ± .00	.53 ± .05	.64 ± .02	.70 ± .01	–
Number of ROIs	5	7	8	7	30	83	68

are highly symmetrical. For standard Desikan atlas hemisphere symmetry is 0.64, and for our best parcellation this value even higher (0.66), and still remains quite high for second level partition (0.55).

Finally we check if our best ensemble parcellation, which combines 400 individual partitions is stable. We split 400 subjects into 2 groups of 200 subjects and independently combine their partitions. We compare resulting parcellations: $\bar{P}_{1,200}$ and $\bar{P}_{201,400}$ between themselves and with original $\bar{P}$ (which is an ensemble of all 400 subjects) again using Adjusted Mutual Information. Both $\bar{P}_{1,200}$ and

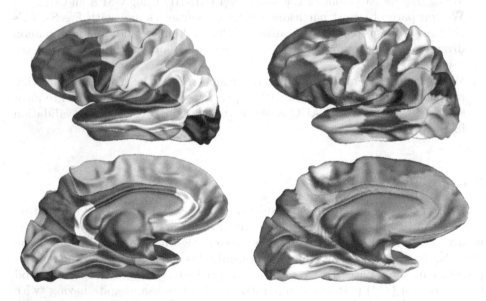

Fig. 2. Left column: Desikan-Killiany parcellation. Right column: HE P^{III} parcellation. Lateral and Medial views, left hemisphere.

$\bar{P}_{201,400}$ shows AMI value greater than 0.80 (0.83 and 0.82 respectively) when compare with $\bar{P}$, they also highly similar between themselves (Fig. 2).

4   Conclusion

We have presented an approach for generating unified connectivity-based human brain atlases bases on consensus clustering. The method is based on finding a pseudo average over the set of individual partitions. Our approach outperforms standard a anatomical parcellation on several important metrics, including agreement with dense connectomes, improved relevance to biological data, and even improved symmetry. Because our approach is entirely data driven an requires no agreement between individual parcellation labels, it combines both the flexibility of individual parcellations and the interpretability of simple unified atlases.

Acknowledgements. This work was funded in part by the Russian Science Foundation grant 17-11-01390 at IITP RAS.

References

1. Blondel, V.D., et al.: Fast unfolding of communities in large networks. J. Stat. Mech. Theory Exp. **2008**(10), P10008 (2008)
2. Dimitriadou, E., Weingessel, A., Hornik, K.: A combination scheme for fuzzy clustering. Int. J. Pattern Recogn. Artif. Intell. **16**(07), 901–912 (2002)
3. Hubert, L., Arabie, P.: Comparing partitions. J. Classif. **2**(1), 193–218 (1985)
4. Kullback, S., Leibler, R.A.: On information and sufficiency. Ann. Math. Stat. **22**(1), 79–86 (1951)
5. Kurmukov, A., et al.: Classifying phenotypes based on the community structure of human brain networks. In: Cardoso, M.J. (ed.) GRAIL/MFCA/MICGen 2017. LNCS, vol. 10551, pp. 3–11. Springer, Cham (2017). https://doi.org/10.1007/978-3-319-67675-3_1
6. Lin, J.: Divergence measures based on Shannon entropy. IEEE Trans. Inf. Theory **37**(14), 145–151 (1991)
7. Meunier, D., Lambiotte, R., Bullmore, E.T.: Modular and hierarchically modular organization of brain networks. Front. Neurosci. **4**, 200 (2010)
8. Moyer, D., et al.: Continuous representations of brain connectivity using spatial point processes. Med. Image Anal. **41**, 32–39 (2017)
9. Nicolini, C., Bordier, C., Bifone, A.: Community detection in weighted brain connectivity networks beyond the resolution limit. Neuroimage **146**, 28–39 (2017)
10. Parisot, S., Glocker, B., Schirmer, M.D., Rueckert, D.: GraMPa: graph-based multimodal parcellation of the cortex using fusion moves. In: Ourselin, S., Joskowicz, L., Sabuncu, M.R., Unal, G., Wells, W. (eds.) MICCAI 2016. LNCS, vol. 9900, pp. 148–156. Springer, Cham (2016). https://doi.org/10.1007/978-3-319-46720-7_18
11. Strehl, A., Ghosh, J.: Cluster ensembles-a knowledge reuse framework for combining multiple partitions. J. Mach. Learn. Res. **3**, 583–617 (2002)
12. Taylor, P.N., Wang, Y., Kaiser, M.: Within brain area tractography suggests local modularity using high resolution connectomics. Sci. Rep. **7**, 39859 (2017)
13. Van Essen, D.C., et al.: The WU-Minn human connectome project: an overview. Neuroimage **80**, 62–79 (2013)

14. Vega-Pons, S., Ruiz-Shulcloper, J.: A survey of clustering ensemble algorithms. Int. J. Pattern Recogn. Artif. Intell. **25**(03), 337–372 (2011)
15. Vinh, N.X., Epps, J., Bailey, J.: Information theoretic measures for clusterings comparison: variants, properties, normalization and correction for chance. J. Mach. Learn. Res. **11**, 2837–2854 (2010)
16. Petrov, D., et al.: Evaluating 35 methods to generate structural connectomes using pairwise classification. arXiv e-prints, eprint = 1706.06031, June 2017
17. Arslan, S., Ktena, S.I., Makropoulos, A., Robinson, E.C., Rueckert, D., Parisot, S.: Human brain mapping: a systematic comparison of parcellation methods for the human cerebral cortex. NeuroImage **170**, 5–30 (2018)

GRAND: Unbiased Connectome Atlas of Brain Network by Groupwise Graph Shrinkage and Network Diffusion

Guorong Wu[1]([✉]), Brent Munsell[2], Paul Laurienti[3], and Moo K. Chung[4]

[1] Department of Psychiatry, University of North Carolina, Chapel Hill, NC 27599, USA
grwu@med.unc.edu
[2] Department of Computer Science, College of Charleston, Charleston, SC 29424, USA
[3] Department of Radiology, Wake Forest School of Medicine, Winston-Salem, NC 27157, USA
[4] Department of Bio, University of Wisconsin, Madison, WI 53706, USA

Abstract. Network science is enhancing our understanding of how the human brain works at a systems level. A complete population-wise mapping of region-to-region connections, called *connectome atlas*, is the key to gaining a more full understanding for network-related brain disorders and for discovering biomarkers for early diagnosis. Since a brain network is commonly encoded in an adjacency matrix, it is difficult to apply the state-of-the-art atlas construction approaches by normalizing and averaging the individual adjacency matrices into a common space. In this paper, we propose a novel data-driven approach to construct an unbiased *connectome atlas* to capture both shared and complementary network topologies across individual brain networks, offering insight into the full spectrum of brain connectivity. Specifically, we employ a hyper-graph to model the manifold of a population of brain networks. In this hypergraph, each node represents the individual participant's brain network, and the edge weight captures the distance between two participants' brain networks. The construction of a *connectome atlas* can be achieved using a hierarchical process of graph shrinkage toward the latent common space where the network topologies of all individual brain networks gradually become similar to each other. During the graph shrinkage, the adjacency matrix of each brain network is transformed to the common space by a series of diffusion matrices which exchange the connectome information with respect to the adjacency matrices on the neighboring hypergraph nodes such that the most representative characteristics of network topology are eventually propagated to the final *connectome atlas*. We have validated our *connectome atlas* construction method on the simulated brain network data and DTI data of 111 twin pairs in determining the genetic contribution of the structural connectivity.

© Springer Nature Switzerland AG 2018
G. Wu et al. (Eds.): CNI 2018, LNCS 11083, pp. 127–135, 2018.
https://doi.org/10.1007/978-3-030-00755-3_14

1 Introduction

The human brain consists of more than 100 billion neurons and 100 trillion connections, making it one of the greatest mysteries and challenges in science. Due to this complexity, the underlying causes of neurological and psychiatric disorders, such as Alzheimer's disease, Parkinson's disease, autism, epilepsy, schizophrenia, and depression are largely unknown. Recent advances in non-invasive and in-vivo neuroimaging technology now allow us to visualize a large-scale map of structural and functional connections in the whole brain at the individual level. The ensemble of macroscopic brain connections can then be described as a complex network - the 'connectome' [1].

Image atlases are critical in neuroimaging studies since they serve as the reference to discover the intrinsic brain differences that are thought to underlie certain neurological disease. In general, the construction process of volumetric (or anatomic) image atlases consists of two major steps [2]: (1) register all individual images into a common space; and (2) average intensities across the warped images at each voxel to produce the atlas representing the common anatomical structures for the entire population.

High inter-subject variations of brain anatomy found in the image atlas construction process suggests that the region-to-region brain connectivity could be even more variable than the brain anatomy itself. Currently, very few methods [3, 4] have addressed the problem of constructing the population-wise *connectome atlas* from a set of brain networks of individual subjects. These works focus on the definition of network nodes, instead of connectivity. Compared to the construction of volumetric image atlases, the computational challenges for *connectome atlases* include the following three issues: **(1)** *The complex nature of network topology.* The region-to-region connections in brain network follow the principle of a "small world" [1], making it much more complex than the fixed neighborhood pattern in 3D regular grid (Fig. 1(a)). **(2)** *Lack of a common geometric space to process the information across brain networks.* Various regularization constraints such as Laplacian and diffeomorphism have been used to convert the ill-posed image registration problem into a well-posed optimization framework [2]. As a result, it is straightforward to deform each voxel until two images become similar. However, it is difficult to "deform" the connectivity of one network to another network with reasonable constraints defined based on the graph data structure. Similarly, averaging brain networks edge-by-edge across individual subjects discounts the multivariate nature of the brain networks, and that issue is not often encountered when averaging individual anatomical images voxel-by-voxel. **(3)** *The difference in network scale, data noise, and acquisition across network datasets.* For instance, thresholding techniques are widely used to discard the spurious connectivity for both structural and functional networks, as shown in Fig. 1(b). Differing network topology based on different thresholds poses a challenge for constructing accurate and robust *connectome atlas*.

To address these challenges, we propose a novel *connectome atlas* construction method to capture both common and varied information across individual brain networks, offering insight into the full spectrum of brain connectivity in the population. Specifically, we first represent the manifold of the brain network data in a hypergraph

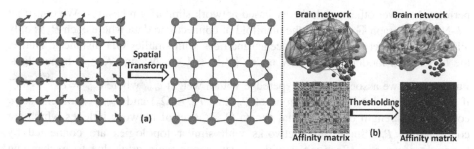

Fig. 1. The challenges for constructing *connectome atlas* compared to volumetric image atlas.

model. Each brain network (encoded in the adjacency matrix) is considered as the hypergraph node. The weight on the hyper-edge measures the distance between two networks. Any two hypergraph nodes are connected via the hyperedge only if the distance between two brain networks is relatively small. Since each node is the individual brain network (a graph) encoded in the adjacency matrix, we use the term "hypergraph" hereafter concerning the manifold of brain network data in the population, for clarity. For each hypergraph node, a non-linear network diffusion procedure is used to exchange the network information through the hyperedges, making the network topology of the underlying hypergraph node similar to the average of its neighboring hypergraph nodes. Essentially, the flow of information exchange occurs in two ways: each brain network is influenced by its neighboring hypergraph nodes, and in turn, the underlying brain network also propagates its native network characteristics to the entire population through the connected hypergraph. This dynamic information exchange procedure continues to converge (to the steady state of diffusion) until the network topologies at all hypergraph nodes become similar to each other. To validate our atlas construction method, we construct the *connectome atlas* on both simulated data and real neuroimaging data. We demonstrate that a more accurate and consistent *connectome atlas* results from our proposed method compared to the existing matrix averaging method [5].

2 Methods

Supposing we have N subjects, each has a brain network encoded as the adjacency matrix P_n $(n = 1,\ldots,N)$ which is derived either from the fiber tractography result on diffusion weighted images or correlation of mean time course in functional MRI data [1]. Each element $p_n(i,j)$ in P_n measures the strength of connectivity between regions O_i and O_j. For simplicity, we assume each P_n is symmetric.

2.1 Model the Manifold of Connectome Data Using a Hypergraph

First, we employ the Gromov-Hausdorff (GH) distance [6] to measure the topological difference between any pair of adjacency matrices P_m and P_n $(m \neq n)$. It has been demonstrated in [6] that GH distance, characterizing the network topology, has superior performance over other graph theory based network similarity measures. We construct a k-NN hypergraph $\mathbb{G}$ to model the manifold of connectome data where each P_n is only allowed to connect to K other adjacency matrices with smaller GH distances. Hence, the hypergraph model consist of N nodes $\{P_n|n = 1, \ldots, N\}$ and $K \times N$ hype-edges. Furthermore, we associate each hyperedge with a weight e_{mn} where $e_{mn} = e^{-\frac{\|GH(P_m,P_n)\|^2}{\sigma^2}}$ if P_m falls in the k-NN neighborhood of P_n (i.e., $P_m \in \Omega_n$) and $e_{mn} = 0$ otherwise. σ controls the strength of exponential decay. Ω_n is a set of network indexes which are connected to P_n. Since only networks with similar topologies are connected by hyperedges, we can effectively avoid a fuzzy *connectome atlas* due to exchanging information too early when two networks have large topological difference.

2.2   Network Diffusion for Individual Brain Networks

To make our method robust to noise and data heterogeneity, we first construct a kernel matrix S_n for each adjacency matrix P_n, where each element $S_n(i,j)$ measures the local adjacency. Suppose ω_n^i denote for the set of region indexes whose connectivity strengths to the underlying region O_i are greater than certain threshold θ. Note, ω_n^i includes the region O_i itself. Thus, we measure the local adjacency as $s_n(i,j) = \frac{p_n(i,j)}{\sum_{k\in\omega_n^i} p_n(i,k)}$ if the index j is in ω_n^i. Otherwise, $s_n(i,j) = 0$.

Pairwise Network Diffusion. Let us first consider the simple case when the hypergraph model $\mathbb{G}$ only consists of two brain networks (P_m and P_n) and one hyperedge between them. The kernel matrix S_m and S_n can be obtained as above. Suppose $P_m^{t=0} = P_m$ and $P_n^{t=0} = P_n$ represent the initial matrix at $t = 0$. We follow the principle of network diffusion in [7] to iteratively update adjacency matrices as:

$$P_m^{t+1} = S_m \times P_n^t \times S_m, \text{ and } P_n^{t+1} = S_n \times P_m^t \times S_n, \tag{1}$$

where P_m^{t+1} and P_n^{t+1} denote for the diffused matrices after t iterations. The intuition of the above network diffusion procedure can be interpreted as $p_m^{t+1}(i,j) = \sum_{u\in\omega_m^i} \sum_{v\in\omega_m^j} s_m(i,u) \times s_m(j,v) \times p_n^t(u,v)$. Thus, the adjacency information is only propagated through the common neighborhood. Suppose regions O_u and O_v are the neighbors of O_i and O_j in the network P_m, respectively. If regions O_u and O_v in the counterpart network P_n also have strong connections, it is highly possible that the connection between O_i and O_j is the most representative characteristic in the population. Another essential fact is that even if O_i and O_j are not strongly connected in P_m, their adjacency still have the chance to remain by propagating $P_n^t(u, v)$ (the presence of their respective neighbors O_u and O_v in the counterpart network P_n) via the network diffusion process. The convergence on pairwise network diffusion has been proven in [7].

Groupwise Network Diffusion. It is straightforward to extend the pairwise network diffusion to the groupwise scenario. Given the hypergraph model $\mathbb{G}$ of connectome data, the diffusion process for each brain network P_m can be defined as:

$$P_m^{t+1} = S_m \times \left(\frac{\sum_{n \in \Omega_m} e_{mn} \cdot P_n^t}{\sum_{n \in \Omega_m} e_{mn}} \right) \times S_m \qquad (2)$$

For computational efficiency, we use the weighted average $\sum_{n \in \Omega_m} e_{mn} \cdot P_n^t$ as the reference in the diffusion process, where larger weight e_{mn} holds more likelihood that the common network topology shared by P_m and P_n can remain in the *connectome atlas*.

2.3 Construction of Connectome Atlas by Groupwise Graph Shrinkage and Network Diffusion (GRAND)

Here, we present our data-driven approach of constructing a *connectome atlas*, as illustrated in Fig. 2. In the beginning ($t = 0$), all brain networks are in their native position, forming the graph model G^0. The lines in Fig. 2 denote the hyperedges where the length of a hyperedge reflects the network distance between the two nodes at either end of the hyperedge. Then, we apply the groupwise network diffusion to each of graph nodes such that the underlying network P_m^t become similar to the weighed average of its connected graph nodes $\sum_{n \in \Omega_m} e_{mn} \cdot P_n$, which makes the hypergraph shrink toward the hidden population center. Through this process all brain networks become similar to each other. The whole procedure of building a *connectome atlas* by GRAND is summarized as follows:

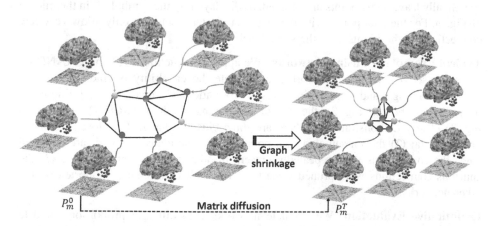

Fig. 2. The overview of our *connectome atlas* construction.

1. Set $t = 0$; calculate the intrinsic matrix S_n for each adjacency matrix P_n;
2. Model the manifold of $\{P_n | n = 1, \ldots, N\}$ with the graph model $\mathbb{G}^0$, where the network distance is calculated by GH distance.
3. For each P_n^t, we apply the network diffusion via Eq. (2) and update to P_n^{t+1};
4. Remodel the manifold of the diffused adjacency matrix $\{P_n^t | n = 1, \ldots, N\}$ with the new graph model $\mathbb{G}^t$;
5. If not converging (e.g., $t < T$, T is the number of total iterations), set $t \leftarrow t + 1$ and go back to step 3; otherwise, stop. Finally, the *connectome atlas* A can be calculated as $A = \sum_{n=1}^{N} P_n^T$.

3   Experiments

We first validate our *connectome atlas* construction method on simulated network data with the known ground truth. Then we evaluate our method on the real brain network data in the cross-sectional twin study since the twin dataset often servers as the ground truth. In all experiments, we compare our GRAND method with log-Euclidian matrix averaging [5] which is abbreviated as log-Euclidean.

3.1   Experiment on Simulated Network Data

Network Simulation. A set of simulated networks were generated from the data shown in the left of Fig. 3, with three separable clusters (displayed in red, blue, and purple). At each iteration, we selectively swap several data points (randomly, near the cluster boundaries) to another cluster. For the data points with correct cluster labels, we construct the connection based on the distance between two points. However, we specifically leave some points un-sampled, as displayed by the symbol '?' in the middle of Fig. 3. For the data points with the incorrect cluster labels, we only allow very few connections to the points with the same label.

Typical Example of Fusing Networks. We apply log-Euclidean [5] and GRAND on the same set of simulated networks, to evaluate the capability of finding common network topology and suppressing noisy information. Compared to log-Euclidean method (blue box in Fig. 3), the network averaging result of GRAND (red box in Fig. 3) clearly demonstrated (**1**) there are three clusters that match closely with the original data; (**2**) the un-sampled data points in one simulated network are captured since the connections were repeatedly present in other network samples; and (**3**) the spurious connections were pruned since they were not consistently supported by the other networks.

Quantitative Evaluation. We simulate networks with different proportions of data swapping percentage, each with 20 simulations. After each simulation, we used log-Euclidean and GRAND methods to construct the average networks. Then we applied the conventional spectral clustering method on the estimated average network to obtain network partitions. We plotted the dice ratio between the clustering for the simulated

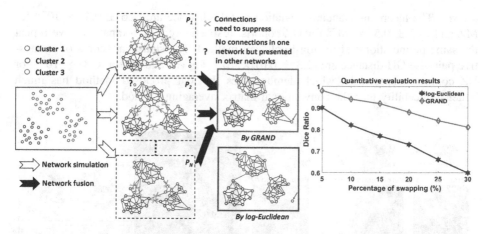

Fig. 3. Average network generated by log-Euclidean (blue box) and GRAND (red box) on the simulated networks (dash boxes) which are drawn from the simulated data (left) with the three distinct clusters (color is used to denote the cluster index). The quantitative evaluation results by log-Euclidean (blue curve) and GRAND (red curve) after network simulation w.r.t. different data swapping percentage is shown in the right. (Color figure online)

ground truth and average networks as a function of the proportion of mis-labeled edges in the right of Fig. 3. The blue and red curves denote the dice ratios that were calculated based on the clustering results derived from the average network by log-Euclidean and GRAND methods, respectively.

3.2   Experiment on Real Brain Network Data

In total 53 dizygotic (DZ) twins and 58 monozygotic (MZ) twins were scanned in a 3.0 T GE scanner with a 32-channel receive-only head coil. Diffusion tensor imaging was performed using a three-shell diffusion-weighted, spin-echo, echo-planar imaging sequence. A total of 6 non-DWI (b = 0) and 63 DWI with non-collinear diffusion encoding directions were collected at b = 500, 800, 2000 (9, 18, 36 directions). Streamline-based tractography was performed and tract counts between AAL parcellations (116 brain regions) were used as adjacency matrices.

Evaluate the Replicability of the Connectome Atlas. We repeated the following steps in DZ twins for 30 permutations: (1) randomly pick one subject out of each DZ twins; (2) form the DZ cohort with 53 individual networks; and (3) construct the *connectome atlas* for the underlying DZ cohort using our GRAND method. We deploy the same procedure to MZ twins. Hence, we can obtain a set of DZ and MZ *connectome atlases* which are shown in the top and bottom of Fig. 4(a), respectively. Through visual inspection, both the DZ and MZ *connectome atlases* consistently have the same network topology across each permutation, which shows the promising replicability performance of our proposed *connectome atlas* construction method. We further quantify the replicability across *connectome atlases* across permutation test by calculating the exhaustive pairwise GH-distance between any two possible *connectome* *atlases.* The mean and standard deviation of GH-distance are $(3.6 \pm 0.4) \times 10^{-3}$ for MZ and $(4.3 \pm 0.5) \times 10^{-3}$ for DZ cohorts, respectively. For comparison, we repeat the same permutation test for log-Euclidean method. The statistical scores of exhaustive pairwise GH-distance are $(2.3 \pm 0.6) \times 10^{-2}$ for MZ and $(3.3 \pm 0.8) \times 10^{-2}$ for DZ cohorts, respectively, which demonstrates that our proposed method has much better replicability than the conventional matrix averaging method.

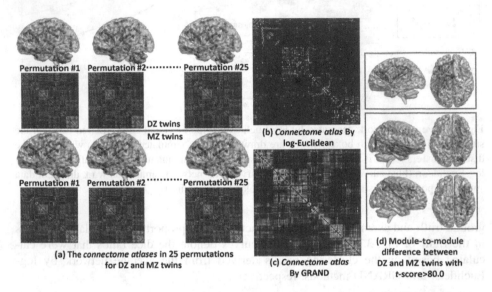

(a) The *connectome atlases* **in 25 permutations for DZ and MZ twins**

(b) *Connectome atlas* By log-Euclidean

(c) *Connectome atlas* By GRAND

(d) Module-to-module difference between DZ and MZ twins with t-score>80.0

Fig. 4. The connectome atlases in totally 25 permutation tests are shown in (a). The connectome atlas by log-Euclidean and GRAND are shown in (b) and (c), respectively. The significance analyses of module-to-module difference between MZ and DZ twins is shown in (d), where we display the top three modules with *t*-score greater than 80.0. (Color figure online)

Discovering the Intrinsic Network Difference Between MZ and DZ Twins. We applied our *connectome atlas* construction method to the entire 53 DZ twin and 58 MZ twin cohorts (in total 222 networks). The *connectome atlases* by log-Euclidean and GRAND are shown in Fig. 4(b) and (c), respectively. It is clear that our *connectome atlas* presents clearer modularity structure than using matrix averaging, where the modularity scores [1] are 0.658 by GRAND and 0.426 by log-Euclidian matrix averaging.

Since MZ twins share 100% of genes while DZ twins only share 50% of genes, it was expected that the distance between MZ twin pairs would be smaller than the distance between DZ twin pair. We first examine the GH-distance for each twin pair based on the networks in their native space and applied a two-sample *t*-test between MZ twin cohort and DZ twin cohort. The *t*-score is only 0.58 ($p > 0.05$), indicating the difference of GH distance is not clearly significant between MZ and DZ cohorts in the native space. Since all brain networks eventually transform to the common space in the end of GRAND, the external difference (not related to gene inheritage) can be substantially reduced in the common space. After two-sample test, we found the GH distance is completely separable between MZ and DZ twins after applying GRAND method (t-score is 79.79 under $p < 10^{-16}$), suggesting that the statistical power has been substantially improved after we transform individual networks to the common space. Furthermore, we examined the significance of module-by-module difference between MZ and DZ twins in the common space based on the modules of *connectome atlas* in Fig. 4(d). Three modules where even the smallest network distance within DZ twins was greater than the largest network distance within MZ twins are shown in the red boxes.

4 Conclusions

We propose a novel computational approach to construct the *connectome atlas* from a set of individual brain networks. We used a hypergraph to model the manifold of individual brain networks. We cast the construction of *connectome atlas* into a dynamic graph shrinking process where the common topology information was propagated along the hypergraph via network diffusion. We achieved promising results on both simulated and real data, which suggests that our connectome atlas methodology will prove valuable to other neuroimaging studies using connectome analyses.

References

1. Sporns, O.: Networks of the Brain. The MIT Press, Cambridge (2011)
2. Joshi, S., Davis, B., Jomier, M., Gerig, G.: Unbiased diffeomorphic atlas construction for computational anatomy. NeuroImage **23**, S151–S160 (2004)
3. Shen, X., Tokoglu, F., Papademetris, X., Constable, R.T.: Groupwise whole-brain parcellation from resting-state fMRI data for network node identification. Neuroimage **82**, 403–415 (2013)
4. Wig, G.S., et al.: Parcellating an individual subject's cortical and subcortical brain structures using snowball sampling of resting-state correlations. Cereb. Cortex **24**, 2036–2054 (2013)
5. Arsigny, V., Fillard, P., Pennec, X., Ayache, N.: Log-Euclidean metrics for fast and simple calculus on diffusion tensors. Magn. Reson. Med. **56**, 411–421 (2006)
6. Lee, H., Chung, M.K., Kang, H., Kim, B.-N., Lee, D.S.: Computing the shape of brain networks using graph filtration and Gromov-Hausdorff metric. In: Fichtinger, G., Martel, A., Peters, T. (eds.) MICCAI 2011. LNCS, vol. 6892, pp. 302–309. Springer, Heidelberg (2011). https://doi.org/10.1007/978-3-642-23629-7_37
7. Wang, B., Jiang, J., Wang, W., Zhou, Z.-h., Tu, Z.: Unsupervised metric fusion by cross diffusion. In: The 25th Conference on Computer Vision and Pattern Recognition, Rhode Island, USA (2012)

Structural Subnetwork Evolution Across the Life-Span: Rich-Club, Feeder, Seeder

Markus D. Schirmer[1,2,3](✉) and Ai Wern Chung[4]

[1] Stroke Division and Massachusetts General Hospital, J. Philip Kistler Stroke Research Center, Harvard Medical School, Boston, MA, USA
mschirmer1@mgh.harvard.edu
[2] Computer Science and Artificial Intelligence Lab,
Massachusetts Institute of Technology, Cambridge, MA, USA
[3] Department of Population Health Sciences, German Centre for Neurodegenerative Diseases (DZNE), Bonn, Germany
[4] Fetal-Neonatal Neuroimaging and Developmental Science Center,
Boston Children's Hospital, Harvard Medical School, Boston, MA, USA

Abstract. The impact of developmental and aging processes on brain connectivity and the connectome has been widely studied. Network theoretical measures and certain topological principles are computed from the entire brain, however there is a need to separate and understand the underlying subnetworks which contribute towards these observed holistic connectomic alterations. One organizational principle is the rich-club - a core subnetwork of brain regions that are strongly connected, forming a high-cost, high-capacity backbone that is critical for effective communication in the network. Investigations primarily focus on its alterations with disease and age. Here, we present a systematic analysis of not only the rich-club, but also other subnetworks derived from this backbone - namely feeder and seeder subnetworks. Our analysis is applied to structural connectomes in a normal cohort from a large, publicly available life-span study. We demonstrate changes in rich-club membership with age alongside a shift in importance from 'peripheral' seeder to feeder subnetworks. Our results show a refinement within the rich-club structure (increase in transitivity and betweenness centrality), as well as increased efficiency in the feeder subnetwork and decreased measures of network integration and segregation in the seeder subnetwork. These results demonstrate the different developmental patterns when analyzing the connectome stratified according to its rich-club and the potential of utilizing this subnetwork analysis to reveal the evolution of brain architectural alterations across the life-span.

Keywords: Connectome · Subnetwork · Life-span · Rich-club
Diffusion

1 Introduction

The last few decades have seen a rapid expansion in the application of network theory to study brain connectivity as it allows for a simple representation of

© Springer Nature Switzerland AG 2018
G. Wu et al. (Eds.): CNI 2018, LNCS 11083, pp. 136–145, 2018.
https://doi.org/10.1007/978-3-030-00755-3_15

complex systems [7,29]. Regions of the brain represent nodes in the network and edges can be defined either structurally or functionally, e.g. by using diffusion MRI (dMRI) or functional MRI (fMRI), respectively. These configurations have been widely studied across individual epochs of development, furthering our understanding of the development and function of the human connectome [3,25, 26,28,30], as well as disease [11,13,23].

The principle that the human connectome is divided up into subnetworks stems from the idea of functional segregation [2,27]. To investigate differences in such subnetworks, studies may utilize a priori divisions of the connectome which are specifically targeted to their research question at hand [5] or determine inter-connections exhibiting a statistical contrast through cluster-based thresholding [33]. However, studies for which such a division cannot be made a priori, data driven measures of modularity may be used, which fraction a connectome into modules [9,18,20]. Others utilize topological features, such as the structural core, to define a subdivision [16]. Whilst the connectome can be interrogated as a whole, identification and analyses of its subnetworks may reveal features with greater regional or network topological specificity and serve as markers of change over time with disease or age.

Many topological aspects of the human brain have been studied, such as small-worldness [32] and economically optimized wiring [8]. Recently, another organizational principle was proposed - the rich-club (RC). The RC is a subset of nodes that is more densely interconnected than expected by chance and has been studied in healthy controls [31], during development [1] and in disease [11,13,23]. Moreover, this definition enables categorization of the edges by their association to the network's 'backbone' that is its rich-club. This stratification identifies edges in the connectome as belonging to: RC, feeder (F) and seeder (S) [1,17,25]. However, this classification has not been propagated to the nodal level to define subnetworks within a connectome, which can subsequently be studied across the life-span.

In this work, we assess the change in properties of subnetworks, as defined by RC, F and S nodes, over the life-span. We first generate the nodal assignment to each category based on group-averaged connectomes for four age groups - younger than 20 years of age, between 20 and 40 years, between 40 and 60 years and above 60 years. We assess these group connectome assignments based on edge density and subsequently propagate the nodal labels back to each subject's connectome. Five subnetworks are defined: RC; F; S; RC and F combined (RC+F); and F and S combined (F+S). For completeness, we include the full connectome in the analyses. Finally, we calculate global network measures of betweenness centrality, efficiency, transitivity and assortativity for each subject and each subnetwork and each subject's connectome and investigate their association with age.

2   Materials and Methods

2.1   Study Design and Patient Population

In this work we utilize data from the NKI/Rockland life-span study [21]. Preprocessed connectome data were obtained from the USC Multimodal Connectivity
database[1] [6]. MRI acquisition details are available elsewhere [6]. In brief, a
total of 196 connectomes of healthy participants are computed, based on 3 T
dMRI acquisition (64 gradient directions; TR, 10000ms; TE, 91ms; voxel size,
$2mm^3$; b-value, 1000 s/mm^2). Following eddy current and motion correction, diffusion tensors are modelled and deterministic tractography was performed, using
fiber assignment by continuous tracking [19] (angular threshold 45°). Regions of
interest (ROI) are based on the Craddock atlas [12], resulting in 188 ROIs and
connections are weighted by the number of streamlines connecting pairs of ROIs.
Here, we normalize each connectome by the maximum streamline count for each
subject, so that the connection weights w_{ij} within each subject are $w_{ij} \in [0, 1]$.

In our analysis, we divide the 196 participants into four age groups: Y20 $\leq$
20 years, 20 years $<$ Y40 $\leq$ 40 years, 40 years $<$ Y60 $\leq$ 60 years, 60 years $<$
Y80 $\leq$ 80 years. Four subjects were above 80 years old (81, 82, 83 and 85 years).
As there were only four subjects, we included them in the Y80 group. Table 1
characterizes the study cohort and groups.

Table 1. NKI/Rockland life-span study cohort characterization and their division by
age (in years).

	Overall	Y20	Y40	Y60	Y80
N	196	53	67	47	29
Age (mean (SD))	35.0 (20.0)	13.41 (4.1)	27.4 (5.9)	47.4 (5.4)	71.0 (6.8)
Sex (Male; %)	58.1	54.7	56.7	72.3	44.8

2.2   Group Connectomes and Rich-Club Organization

A group-averaged connectome computed from weighted matrices is derived in
two steps [31]. First, we calculate a binarized, group-average adjacency matrix
by retaining edges that are present in at least 90% of the subjects in each group.
Weights are subsequently added to the group-averaged adjacency matrix by taking the average weight of each connection across the group, generating a weighted
group-averaged connectome W_{group} for each age group.

We utilize W_{group} to subsequently calculate the weighted RC parameter
$\phi_{group}(k)$ [22] as implemented in the Brain Connectivity Toolbox [24], where
k denotes the degree of nodes. The RC parameter $\phi_{group}(k)$ is normalized relative to a set of comparable random networks of equal size and with similar

---

[1] http://umcd.humanconnectomeproject.org.

connectivity distribution. Here, we generate 100 random networks while preserving weight, degree and strength distributions of W_{group} [24]. For each of these random realizations of the graph, we calculate the weighted RC parameter $\phi_{rand}(k)$. Finally, the normalized weighted RC parameter is calculated as

$$\phi_{group}^{norm}(k) = \frac{\phi_{group}^{norm}(k)}{mean(\phi_{rand}(k))}. \tag{1}$$

For this metric, $\phi_{group}^{norm}(k) > 1$ denotes the presence of a RC. We select

$$k_{max}^{group} : \phi_{group}^{norm}(k) > 1, \tag{2}$$

as the degree of the RC nodes of a given group, which allows us to determine the RC members. This analysis is repeated for each group and k_{max}^{group} is recorded. We then use the mean of k_{max}^{group} over all groups for all subsequent analyses.

2.3 Subnetwork Definition

In previous work, edges have often been differentiated after RC assessment, stratified according to their relation to the RC nodes - RC edges connect RC nodes; F edges connect RC to non-RC nodes; and S edges connect any two non-RC nodes [1,17,25]. Similarly, nodes can be differentiated into RC, F and S, based on their connectivity profile. F nodes are directly connected to RC nodes, but do not belong to the RC themselves. S nodes do not belong to the RC nor are they directly connected to an RC node. Here, we define five subnetworks using this differentiation, namely RC alone (RC), F alone (F), S alone (S), as well as the subnetworks from combinations of RC, F and S, including the connections between them (RC+F; F+S; RC+F+S). These subnetworks are identified for each group using the group-averaged connectome and subsequently propagated to each subject. Figure 1 illustrates a toy network example of and it's separation into RC, F, and S, with an example connectome for Y20.

2.4 Network Analysis

There are a variety of network measures which can be derived from any given graph or subnetwork, describing different local or global properties [10,24,31,32]. In this work we focus on global network measures of betweenness centrality (BC), global efficiency (E), transitivity (T) and assortativity (a). BC relates to the amount of information passing through a given node, thereby reflecting its importance for communication in the network. Although it is a local measure, the mean BC describes the prevalence of important nodes in a network. E characterizes how efficiently a network exchanges information. It also directly relates to the global integration of the network, where higher E reflects greater integration between specialized communities. T describes the likelihood of closed triangles in a network. If node n is connected to node m, which in turn is connected to node o, then T reflects the probability that node n is also directly connected to node

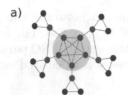 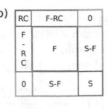 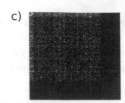

Fig. 1. Differentiation of RC, F and S nodes. (a) Toy model of a network, showing the highly connected RC center (purple), F nodes (blue) with direct connections to the RC, but are not RC members themselves, and S nodes (black) which are not RC members nor connected to the RC. (b) Regions within a connectivity matrix, classified by their membership to RC, F, or S nodes, as well as the connectivity between memberships (green, orange and gray). (c) Connectome of Y20 (=20 years) group, rearranged to reflect the representation in b). Color-coded to represent RC (white), F (yellow) and S (red) connections. (Color figure online)

o. These closed triangles lead to locally segregated networks, thereby promoting specialization [2,27]. The last measure, a, is a correlation coefficient between the edge weights of node on opposite ends of a connection. If a is positive, it means that the connected nodes have a similar connectivity profile, reflecting a tendency for similar nodes in a network to be linked. Importantly, these measures can be calculated in networks with multiple, disconnected components. After investigating the group differences of these four network measures, we calculate Spearman's correlation coefficient for each network measure against age for the entire NKI/Rockland cohort in each of the four groups to identify developmental trends. For the group-averaged connectomes, we further assess the edge density of the adjacency matrices for each of the five subnetworks.

3   Results

3.1   Group Connectome, Rich-Club and Subnetwork Definition

Table 2 details the k_{max}^{group} computed from each group-averaged connectome and the number of corresponding regions determined to form the rich-club subnetwork. The average k_{max}^{group} over all groups quantifying the common degree for RC assessment is $k_{max} = 55.25$.

Table 2. Normalized RC analysis of the four age groups, identifying the degree k_{max}^{group}, as well as the corresponding number of regions.

	Y20	Y40	Y60	Y80
k_{max}^{group}	56	54	55	56
# regions	8	11	10	7

Using k_{max} results in the identification of ten individual regions as belonging to the rich-club across the life-span cohort, with each region appearing in at least two of the four groups in our analysis (see Fig. 2b). For each age group, Fig. 2a shows the corresponding connectomes reordered by their relation to the rich-club, Fig. 2b) lists the regions that are determined to be part of the RC for each group, and Fig. 2c) shows the edge densities within and between nodal membership on the group-averaged connectomes.

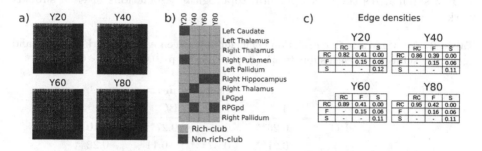

a) Y20 Y40 Y60 Y80

b)
Left Caudate
Left Thalamus
Right Thalamus
Right Putamen
Left Pallidum
Right Hippocampus
Right Thalamus
LPGpd
RPGpd
Right Pallidum

Rich-club
Non-rich-club

c) Edge densities

Y20

	RC	F	S
RC	0.82	0.41	0.00
F	-	0.15	0.05
S	-	-	0.12

Y40

	RC	F	S
RC	0.86	0.39	0.00
F	-	0.15	0.06
S	-	-	0.11

Y60

	RC	F	S
RC	0.89	0.41	0.00
F	-	0.15	0.06
S	-	-	0.11

Y80

	RC	F	S
RC	0.95	0.42	0.00
F	-	0.16	0.06
S	-	-	0.11

Fig. 2. Membership analysis of the four age groups. (a) Group-averaged connectomes for each group reordered as in Fig. 1b, color-coded to represent RC (white), F (yellow) and S (red) connections. (b) Brain regions identified as belonging to the RC for each of the four group-averaged connectomes (L/RPGpd: Left/Right Parahippocampal posterior). (c) Edge density analysis in each section of the connectivity matrix (see Fig. 1b) after nodal assignment to either RC, F, or S. (Color figure online)

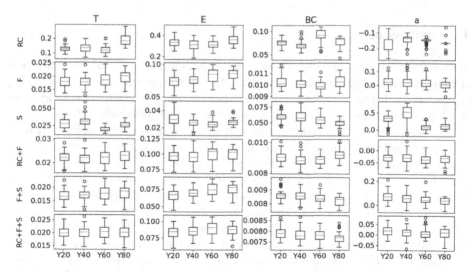

Fig. 3. Network theoretical measures by subnetworks for each age group.

3.2   Subnetwork Network Analysis

Following the RC, F, S differentiation, we label each subject's nodes with the corresponding assignment based on their respective group-averaged connectome. Figure 3 shows the distributions of mean values for each network theoretical measure computed from each subnetwork in each age group.

Utilizing the age of all 196 subjects in the cohort, we calculate Spearman's rank correlation coefficient for each network measure within each subnetwork. Table 3 summarizes the developmental topological associations of each subnetwork with age.

Table 3. Spearman's rank correlation coefficients between each network measure and age of the subjects for each subnetwork (*: $p<0.05$; **: $p<0.01$; ***: $p<0.001$).

		Network measure			
		T	E	BC	a
Subnetwork	RC	0.26***	0.05	0.27***	−0.10
	F	0.21**	0.37***	−0.11	−0.22**
	S	−0.29***	−0.21**	−0.4***	−0.46***
	RC+F	0.09	0.19**	0.03	−0.13
	F+S	0.18**	0.37***	−0.33***	−0.34***
	RC+F+S	0.06	0.2**	−0.25***	−0.2**

4   Discussion

In this work we presented a subnetwork analysis based on the differentiation of rich-club, feeder and seeder regions in the brain over the life-span. Utilizing group-averaged connectomes for subnetwork definition, we were able to identify developmental patterns in network measures from each subnetwork. Importantly, we demonstrated that rich-club, feeder and seeder regions evolve differently over time, potentially reflecting their functional difference in the human connectome.

By dividing our cohort into individual groups based on their age, we showed that the determined degree for the presence of a RC organization is largely consistent across age ranges. However, the number of regions identified as RC varied. We stabilized the number of RC regions by employing the average degree for our subnetwork analyses. In their study, van den Heuvel et al. [31] reported six bilateral regions (precuneus, superior frontal, superior parietal cortex, hippocampus, putamen and thalamus) making up the adult human connectome's RC. We note that our analyses comprised of only subcortical regions in the rich-club. While we were not able to reproduce the cortical RC regions found by van den Heuvel et al., our results agree with the remaining three subcortical RC regions (hippocampus, putamen and thalamus). In addition, we identified left caudate, and bilateral pallidum and parahippocampal posterior as part of the RC.

Considering the variation across the age groups and the use of weighted connectomes, it is possible that RC membership may be a more fluid process, where specific regions may gain or lose their membership depending on age-specific requirements of the brain over the life-span (e.g. reflecting the neurobiological, metabolic and cognitive demands to meet developmental and aging processes). As such, others have similarly found this fluidity in rich-club membership in comparisons of children [15] and adolescents [14] versus young adults. This is likely an effect of changing connectivity weights, as the edge density remains mostly constant for F, S, F-S and RC-F connections. It is only within the RC subnetwork that density increases with age.

Analyzing network measures for each of the defined subnetworks demonstrates varying associations with age for each of the subnetworks. Our results show that the relationship between network measures with age is not linear, which is in agreement with other studies [4,34]. However, we observe clear trends for T and BC within the RC with age, suggesting that the RC may be further reinforced (increase in T) and that members of the RC might not be of equal importance in terms of information transport (increasing BC). Feeder regions are further integrated in the network, increasing the E of their connectivity with age, while further strengthening their communication paths within this subnetwork. Seeder, however, seemingly reduce in importance with age, shown in a reduction of every network measure investigated with age. The reduced importance may be indicative of a brain reorganization or pruning-like process, where increasingly limited resources for the human connectome are invested in the feeder regions.

There are limitations to our study. The identification of RC regions, though relatively stable across age groups, varies. This may be due to underlying biological processes, however, further investigation is necessary, in order to ensure that this is not the result of noise in the data or due to the cross-sectional nature of the study cohort. Furthermore, mostly subcortical regions were identified as being part of the RC. Future studies can use further divisions of the human connectome, e.g. into cortico-cortical and/or subcortico-subcortical connectomes, to elucidate more specific patterns of age-related change in the cortex. In addition, the use of multi-modal MRI data can further enhance our understanding of developmental and aging trajectories. Another important aspect will therefore be the use of combined structural and functional connectome data for analysis.

In this study we report different patterns of evolution in subnetworks defined using rich-club topology over the life-span. We demonstrate in particular, that feeder regions are strengthened, while the seeder subnetwork is weakened with age. Future, multi-modal studies in healthy controls will allow the formulation of novel hypotheses, which can subsequently be tested in disease, and have the potential of identifying biomarkers for diagnoses and prognoses across a breadth of neuropathological and neurodevelopmental disorders.

Funding. This project has received funding from the European Union's Horizon 2020 research and innovation programme under the Marie Sklodowska-Curie grant agreement No 753896.

References

1. Ball, G., et al.: Rich-club organization of the newborn human brain. Proc. Nat. Acad. Sci. **111**(20), 7456–7461 (2014)
2. Bassett, D.S., Meyer-Lindenberg, A., Achard, S., Duke, T., Bullmore, E.: Adaptive reconfiguration of fractal small-world human brain functional networks. Proc. Nat. Acad. Sci. **103**(51), 19518–19523 (2006)
3. Batalle, D., Edwards, A.D., O'Muircheartaigh, J.: Annual research review: not just a small adult brain: understanding later neurodevelopment through imaging the neonatal brain. J. Child Psychol. Psychiatry **59**(4), 350–371 (2018)
4. Batalle, D., et al.: Early development of structural networks and the impact of prematurity on brain connectivity. NeuroImage **149**, 379–392 (2017)
5. Bonilha, L., et al.: Presurgical connectome and postsurgical seizure control in temporal lobe epilepsy. Neurology **81**(19), 1704–1710 (2013)
6. Brown, J.A., Rudie, J.D., Bandrowski, A., Van Horn, J.D., Bookheimer, S.Y.: The UCLA multimodal connectivity database: a web-based platform for brain connectivity matrix sharing and analysis. Front. Neuroinf. **6**, 28 (2012)
7. Bullmore, E., Sporns, O.: Complex brain networks: graph theoretical analysis of structural and functional systems. Nat. Rev. Neurosci. **10**(3), 186 (2009)
8. Bullmore, E., Sporns, O.: The economy of brain network organization. Nat. Rev. Neurosci. **13**(5), 336 (2012)
9. Cao, M., et al.: Topological organization of the human brain functional connectome across the lifespan. Dev. Cogn. Neurosci. **7**, 76–93 (2014)
10. Chung, A.W., et al.: Characterising brain network topologies: a dynamic analysis approach using heat kernels. Neuroimage **141**, 490–501 (2016)
11. Collin, G., Kahn, R.S., de Reus, M.A., Cahn, W., van den Heuvel, M.P.: Impaired rich club connectivity in unaffected siblings of schizophrenia patients. Schizophrenia Bull. **40**(2), 438–448 (2013)
12. Craddock, R.C., James, G.A., Holtzheimer III, P.E., Hu, X.P., Mayberg, H.S.: A whole brain fMRI atlas generated via spatially constrained spectral clustering. Hum. Brain Mapp. **33**(8), 1914–1928 (2012)
13. Daianu, M.: Rich club analysis in the Alzheimer's disease connectome reveals a relatively undisturbed structural core network. Hum. Brain Mapp. **36**(8), 3087–3103 (2015)
14. Dennis, E.L., et al.: Development of the rich club in brain connectivity networks from 438 adolescents & adults aged 12 to 30. In: Proceedings of IEEE International Symposium Biomed Imaging, pp. 624–627 (2013)
15. Grayson, D.S., et al.: Structural and functional rich club organization of the brain in children and adults. PLOS ONE **9**(2), e88297 (2014)
16. Hagmann, P., et al.: Mapping the structural core of human cerebral cortex. PLoS Biol. **6**(7), e159 (2008)
17. van den Heuvel, M.P., Kahn, R.S., Goñi, J., Sporns, O.: High-cost, high-capacity backbone for global brain communication. Proc. Nat. Acad. Sci. **109**(28), 11372–11377 (2012)
18. Ingalhalikar, M., et al.: Sex differences in the structural connectome of the human brain. Proc. Nat. Acad. Sci. **111**(2), 823–828 (2014)
19. Mori, S., Crain, B.J., Chacko, V.P., Van Zijl, P.C.: Three-dimensional tracking of axonal projections in the brain by magnetic resonance imaging. Ann. Neurol. Official J. Am. Neurol. Assoc. Child Neurol. Soc. **45**(2), 265–269 (1999)

20. Moussa, M.N., Steen, M.R., Laurienti, P.J., Hayasaka, S.: Consistency of network modules in resting-state fMRI connectome data. PloS One **7**(8), e44428 (2012)
21. Nooner, K.B., et al.: The NKI-rockland sample: a model for accelerating the pace of discovery science in psychiatry. Front. Neurosci. **6**, 152 (2012)
22. Opsahl, T., Colizza, V., Panzarasa, P., Ramasco, J.J.: Prominence and control: the weighted rich-club effect. Physical Rev. Lett. **101**(16), 168702 (2008)
23. Ray, S., et al.: Structural and functional connectivity of the human brain in autism spectrum disorders and attention-deficit/hyperactivity disorder: a rich club-organization study. Hum. Brain Mapp. **35**(12), 6032–6048 (2014)
24. Rubinov, M., Sporns, O.: Complex network measures of brain connectivity: uses and interpretations. Neuroimage **52**(3), 1059–1069 (2010)
25. Schirmer, M.D.: Developing brain connectivity: effects of parcellation scale on network analysis in neonates. Ph.D. thesis, King's College London (2015)
26. Sporns, O.: Networks of the Brain. MIT press, Cambridge (2010)
27. Sporns, O.: Network attributes for segregation and integration in the human brain. Curr. Opin. Neurobiol. **23**(2), 162–171 (2013)
28. Sporns, O., Chialvo, D.R., Kaiser, M., Hilgetag, C.C.: Organization, development and function of complex brain networks. Trends Cogn. Sci. **8**(9), 418–425 (2004)
29. Sporns, O., Tononi, G., Kötter, R.: The human connectome: a structural description of the human brain. PLoS Comput. Biol. **1**(4), e42 (2005)
30. Tymofiyeva, O., et al.: Towards the baby connectome: mapping the structural connectivity of the newborn brain. PloS One **7**(2), e31029 (2012)
31. Van Den Heuvel, M.P., Sporns, O.: Rich-club organization of the human connectome. J. Neurosci. **31**(44), 15775–15786 (2011)
32. Watts, D.J., Strogatz, S.H.: Collective dynamics of small-world networks. Nature **393**(6684), 440 (1998)
33. Zalesky, A., Fornito, A., Bullmore, E.T.: Network-based statistic: identifying differences in brain networks. Neuroimage **53**(4), 1197–1207 (2010)
34. Zhao, T., et al.: Age-related changes in the topological organization of the white matter structural connectome across the human lifespan. Hum. Brain Mapp. **36**(10), 3777–3792 (2015)

26. Muldoon, S. F., et al.: Stimulation-based control of dynamic brain network. PLoS Comput. Biol. **12**(9), e1005076 (2016)

27. Betzel, R. F., et al.: Optimally controlling the human connectome: a network analysis approach. Sci. Rep. **6**, 30770 (2016)

28. Gu, S., et al.: Controllability of structural brain networks. Nat. Commun. **6**, 8414 (2015)

29. Tang, E., et al.: Developmental increases in white matter network controllability support a growing diversity of brain dynamics. Nat. Commun. **8**(1), 1252 (2017)

30. ...

...

Author Index

Printed in the United States
By Bookmasters